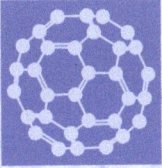

nanomaterials

Electrochemically Engineering of Nanoporous Materials

Edited by
Abel Santos

Printed Edition of the Special Issue Published in *Nanomaterials*

www.mdpi.com/journal/nanomaterials

MDPI

Electrochemically Engineering of Nanoporous Materials

Electrochemically Engineering of Nanoporous Materials

Special Issue Editor

Abel Santos

MDPI • Basel • Beijing • Wuhan • Barcelona • Belgrade

MDPI

Special Issue Editor
Abel Santos
The University of Adelaide
Australia

Editorial Office
MDPI
St. Alban-Anlage 66
Basel, Switzerland

This is a reprint of articles from the Special Issue published online in the open access journal *Nanomaterials* (ISSN 2079-4991) from 2017 to 2018 (available at: http://www.mdpi.com/ journal/nanomaterials/special_issues/electrochem_eng_nanoporous_mater)

For citation purposes, cite each article independently as indicated on the article page online and as indicated below:

LastName, A.A.; LastName, B.B.; LastName, C.C. Article Title. *Journal Name* **Year**, *Article Number*, Page Range.

ISBN 978-3-03897-268-6 (Pbk)
ISBN 978-3-03897-269-3 (PDF)

Cover image courtesy of Tom Macdonald and Steven J.P McInnes.

Contents

About the Special Issue Editor

Abel Santos, Dr. I received my BEng in Chemical Engineering from the Universitat Jaume I (Spain) in 2006, and MEng and PhD degrees in Electronic Engineering from the Universitat Rovira i Virgili (Spain) in 2007 and 2011, respectively. In 2014, I was awarded an ARCDECRA, and currently I am a 'Research for Impact' Fellow in the School of Chemical Engineering, a member of the Institute for Photonics and Advanced Sensing (IPAS), and an associate investigator of the ARC Centre of Excellence for Nanoscale BioPhotonics (CNBP) at the University of Adelaide, where I work on the structural engineering of photonic crystals based on nanoporous anodic alumina.

Preface to "Electrochemically Engineering of Nanoporous Materials"

Nanoporous materials are outstanding platforms due to their unique chemical and physical properties at the nanoscale, which make them suitable candidates to develop advanced materials and systems for a plethora of applications, including catalysis and photocatalysis, energy harvesting and storage, photonics and optoelectronics, nanomedicine, and filtration and separation. Among different methods, electrochemical fabrication techniques offer many advantages over conventional nanofabrication methods to produce nanoporous materials with precisely engineered properties such as controllability, reproducibility, high resolution, scalability, high-throughput, cost-competitiveness, and time-efficient processes. Despite numerous advances in this area, electrochemical engineering of nanoporous materials is a highly dynamic and broad research field that is constantly enabling excellent opportunities for further trans-disciplinary fundamental and applied research.

In this context, this Special Issue of *Nanomaterials* compiles a series of illustrative examples on several fundamental aspects and inter-disciplinary applications of nanoporous materials produced by different electrochemical and chemical methods, from energy to drug delivery. It is thus expected that the field of electrochemically engineered nanoporous materials will continue to grow and spread towards more sophisticated applications. Metallic and semiconductor nanoporous materials can enable the precise control of light-matter interactions such as surface plasmon resonance, photonic crystal, and slow photon effect at the nanoscale for optical sensing and biosensing, energy harvesting, and environmental remediation applications. Nanoporous materials with well-defined nanostructures enable new opportunities to study molecular interactions to develop advanced materials with unique chemical and physical properties for ultra-efficient separation and filtration processes such as water desalination. Nanoporous materials based on inert and non-cytotoxic materials provide an excellent matrix to load and accommodate therapeutics, which can be passively or actively released by local or remote triggers in a timely fashion for personalised medical therapies. These structures also provide unique template platforms for the synthesis of other nanostructures with controlled geometric features.

Abel Santos
Special Issue Editor

nanomaterials

MDPI

Editorial

Electrochemical Engineering of Nanoporous Materials

Abel Santos [1,2,3,*]

[1] School of Chemical Engineering, The University of Adelaide, Adelaide, SA 5005, Australia
[2] Institute for Photonics and Advanced Sensing (IPAS), The University of Adelaide, Adelaide, SA 5005, Australia
[3] ARC Centre of Excellence for Nanoscale BioPhotonics (CNBP), The University of Adelaide, Adelaide, SA 5005, Australia
* Correspondence: abel.santos@adelaide.edu.au; Tel.: +61-8-8313-1535

Received: 22 August 2018; Accepted: 5 September 2018; Published: 6 September 2018

Nanoporous materials are outstanding platforms due to their unique chemical and physical properties at the nanoscale, which make them suitable candidates to develop advanced materials and systems for a plethora of applications, including catalysis and photocatalysis [1–7], energy harvesting and storage [8], photonics and optoelectronics [9,10], nanomedicine [11–15], and filtration and separation [16–18]. Among different methods, electrochemical fabrication techniques offer many advantages over conventional nanofabrication methods to produce nanoporous materials with precisely engineered properties, such as controllability, reproducibility, high resolution, scalability, high-throughput, cost-competitiveness, and time-efficient processes. Despite numerous advances in this area, electrochemical engineering of nanoporous materials remains a highly dynamic and broad research field that continues to enable excellent opportunities for further trans-disciplinary fundamental and applied research.

In this context, this Special Issue of *Nanomaterials* collates a series of illustrative examples on several fundamental aspects and inter-disciplinary applications of nanoporous materials produced by different electrochemical and chemical methods, from energy to drug delivery. It is thus expected that the field of electrochemically engineered nanoporous materials will continue to grow and spread towards more sophisticated applications. Metallic and semiconductor nanoporous materials can enable the precise control of light–matter interactions, such as surface plasmon resonance, photonic crystal and slow photon effect at the nanoscale for optical sensing and biosensing, energy harvesting, and environmental remediation applications. Nanoporous materials with well-defined nanostructures enable new opportunities to study molecular interactions to develop advanced materials with unique chemical and physical properties for ultra-efficient separation and filtration processes, such as water desalination. Nanoporous materials based on inert and non-cytotoxic materials provide an excellent matrix to load and accommodate therapeutics, which can be passively or actively released by local or remote triggers in a timely fashion for personalised medical therapies. These structures also provide unique template platforms for the synthesis of other nanostructures with controlled geometric features.

Finally, I would like to thank all authors of this Special Issue of *Nanomaterials* for their contributions and all referees for their valuable comments, suggestions and time, as well as the editorial office for their constant and swift support throughout.

Funding: The author thanks the support provided by the Australian Research Council (ARC) through the grant number CE140100003, the School of Chemical Engineering, the University of Adelaide (DVCR 'Research for Impact' initiative), the Institute for Photonics and Advanced Sensing (IPAS), and the ARC Centre of Excellence for Nanoscale BioPhotonics (CNBP).

Conflicts of Interest: The author declares no conflict of interest.

References

1. Hu, Y.S.; Guo, Y.G.; Single, W.; Hore, S.; Balaya, P.; Maier, J. Electrochemical lithiation synthesis of nanoporous materials with superior catalytic and capacitive activity. *Nat. Mater.* **2006**, *5*, 713–717. [CrossRef] [PubMed]
2. Xu, W.; Zhu, S.; Lian, Y.; Li, Z.; Cui, Z.; Yang, X.; Inoue, A. Nanoporous CuS with excellent photocatalytic property. *Sci. Rep.* **2015**, *5*, 18125. [CrossRef] [PubMed]
3. An, H.R.; Park, S.Y.; Kim, H.; Lee, C.Y.; Choi, S.; Lee, S.C.; Seo, S.; Park, E.C.; Oh, Y.K.; Song, C.G.; et al. Advanced nanoporous TiO$_2$ photocatalysts by hydrogen plasma for efficient solar-light photocatalytic application. *Sci. Rep.* **2016**, *6*, 29683. [CrossRef] [PubMed]
4. Geng, Z.; Zhang, Y.; Yuan, X.; Huo, M.; Zhao, Y.; Lu, Y.; Qiu, Y. Incorporation of Cu$_2$O nanocrystals into TiO$_2$ photonic crystal for enhanced UV-visible light driven photocatalysis. *J. Alloys Compd.* **2015**, *644*, 734–741. [CrossRef]
5. Lu, Y.; Yu, H.; Chen, S.; Quan, X.; Zhao, H. Integrating plasmonic nanoparticles with TiO$_2$ photonic crystal for enhancement of visible-light driven photocatalysis. *Environ. Sci. Technol.* **2012**, *46*, 1724–1730. [CrossRef] [PubMed]
6. Sheng, X.; Liu, J.; Coronel, N.; Agarwal, A.M.; Michel, J.; Kimerling, L.C. Integration of self-assembled porous alumina and distributed Bragg reflector for light trapping in Si photovoltaic devices. *IEEE Photonics Technol. Lett.* **2010**, *22*, 1394–1396. [CrossRef]
7. Guo, M.; Xie, K.; Wang, Y.; Zhou, L.; Huang, H. Aperiodic TiO$_2$ nanotube photonic crystal: Full-visible-spectrum solar light harvesting in photovoltaic devices. *Sci. Rep.* **2014**, *4*, 6442. [CrossRef] [PubMed]
8. Li, Y.; Fu, Z.Y.; Su, B.L. Hierarchically structured porous materials for energy conversion and storage. *Adv. Funct. Mater.* **2012**, *22*, 4634–4667. [CrossRef]
9. Zheng, X.; Meng, S.; Chen, J.; Wang, J.; Xian, J.; Shao, Y.; Fu, X.; Li, D. Titanium dioxide photonic crystals with enhanced photocatalytic activity: Matching photonic band gaps of TiO$_2$ to the absorption peaks of dyes. *J. Phys. Chem. C* **2013**, *117*, 21263–21273. [CrossRef]
10. Santos, A.; Balderrama, V.S.; Alba, M.; Formentín, P.; Ferré-Borrull, J.; Pallarès, J.; Marsal, L.F. Nanoporous anodic alumina barcodes: Toward smart optical biosensors. *Adv. Mater.* **2012**, *24*, 1050–1054. [CrossRef] [PubMed]
11. Jeon, G.; Yang, S.Y.; Byun, J.; Kim, J.K. Electrically actuatable smart nanoporous membrane for pulsatile drug release. *Nano Lett.* **2011**, *11*, 1284–1288. [CrossRef] [PubMed]
12. Santos, A.; Aw, M.S.; Bariana, M.; Kumeria, T.; Wang, Y.; Losic, D. Drug-releasing implants: Current progress, challenges and perspective. *J. Mater. Chem. B* **2014**, *2*, 6157–6182. [CrossRef]
13. Anglin, E.J.; Schwartz, M.P.; Ng, V.P.; Perelman, L.A.; Sailor, M.J. Engineering the chemistry and nanostructure of porous silicon Fabry-Pérot films for loading and release of a steroid. *Langmuir* **2004**, *20*, 11264–11269. [CrossRef] [PubMed]
14. Wu, E.C.; Andrew, J.S.; Cheng, L.; Freeman, W.R.; Pearson, L.; Sailor, M.J. Real-time monitoring of sustained drug release using the optical properties of porous silicon photonic crystal particles. *Biomaterials* **2011**, *32*, 1957–1966. [CrossRef] [PubMed]
15. Low, S.P.; Williams, K.A.; Canham, L.T.; Voelcker, N.H. Evaluation of mammalian cell adhesion on surface-modified porous silicon. *Biomaterials* **2006**, *27*, 4538–4546. [CrossRef] [PubMed]
16. Burham, N.; Hamzah, A.A.; Yunas, J.; Majlis, B.Y. Electrochemically etched nanoporous silicon membrane for separation of biological molecules in mixture. *J. Micromech. Microeng.* **2017**, *27*, 075021. [CrossRef]
17. Alsawat, M.; Altalhi, T.; Santos, A.; Losic, D. Carbon nanotubes-nanoporous anodic alumina composite membranes: Influence of template on structural, chemical and transport properties. *J. Phys. Chem. C* **2018**, *121*, 13634–13644. [CrossRef]
18. Altalhi, T.; Kumeria, T.; Santos, A.; Losic, D. Synthesis of well-organised carbon nanotube membranes for tuneable ionic transport from non-degradable commercial plastic bags: Towards nanotechnological recycling. *Carbon* **2013**, *63*, 423–433. [CrossRef]

nanomaterials

MDPI

Article

Single-Walled Carbon Nanotube (SWCNT) Loaded Porous Reticulated Vitreous Carbon (RVC) Electrodes Used in a Capacitive Deionization (CDI) Cell for Effective Desalination

Ali Aldalbahi [1,*], **Mostafizur Rahaman** [1], **Mohammed Almoiqli** [2], **Abdelrazig Hamedelniel** [1] and **Abdulaziz Alrehaili** [1]

[1] Department of Chemistry, College of Science, King Saud University, Riyadh 11451, Saudi Arabia; mrahaman@ksu.edu.sa (M.R.); aelfaki@ksu.edu.sa (A.H.); 436107406@student.ksu.edu.sa (A.A.)
[2] Nuclear Sciences Research Institute, King Abdulaziz City for Science and Technology, Riyadh 11442, Saudi Arabia; almoiqli@kacst.edu.sa
* Correspondence: aaldalbahi@ksu.edu.sa; Tel.: +966-114-675-883

Received: 7 March 2018; Accepted: 3 July 2018; Published: 13 July 2018

Abstract: Acid-functionalized single-walled carbon nanotube (a-SWCNT)-coated reticulated vitreous carbon (RVC) composite electrodes have been prepared and the use of these electrodes in capacitive deionization (CDI) cells for water desalination has been the focus of this study. The performance of these electrodes was tested based on the applied voltage, flow rate, bias potential and a-SWCNT loadings, and then evaluated by electrosorption dynamics. The effect of the feed stream directly through the electrodes, between the electrodes, and the distance between the electrodes in the CDI system on the performance of the electrodes has been investigated. The interaction of ions with the electrodes was tested through Langmuir and Freundlich isotherm models. A new CDI cell was developed, which shows an increase of 23.96% in electrosorption capacity compared to the basic CDI cells. Moreover, a comparison of our results with the published results reveals that RVC/a-SWCNT electrodes produce 16 times more pure water compared to the ones produced using only CNT-based electrodes. Finally, it can be inferred that RVC/a-SWCNT composite electrodes in newly-developed CDI cells can be effectively used in desalination technology for water purification.

Keywords: RVC/a-SWCNT composite electrode; capacitive deionization; water purification; desalination technology

---

1. Introduction

Capacitive deionization (CDI) technology has become a new desalination technology developed in recent years with advantages such as energy saving, no secondary pollution, and simple operation, showing a broad prospect that this low-pressure and non-membrane desalination technology employs the basic electrochemical principle of adsorbing ions to high surface-area electrodes in a capacitive fashion such that the outgoing stream becomes devoid of the ions that were present in the incoming stream. The realization of high removal efficiency for capacitive deionization is strongly correlated with the high capacitance value, large surface area for ion accumulation, good electrical conductivity, and suitable pore size of the electrodes. To date, extensive studies have been made on carbon materials that are used as CDI electrodes [1–12]. Moreover, there are also studies based on the use of carbon nanotubes (CNTs) as CDI electrodes because of their exceptionally affordable functionality, structural stability, high surface area, and low electrical resistivity [13–16].

In a study, CNTs were used as electrodes in CDI systems, where the data obtained through experiments agree well when compared with the Langmuir model, and the highest desalination capacity was found to be 40 mg/g [17]. The desalination capacity was improved by compositing CNTs with polyaniline (PAni) [18]. The electrosorption capacity of CNTs/PAni electrodes was found to be higher than the electrode made from only SWCNTs, where the electrodes based on composite materials were regenerated, indicating excellent recyclability. CNT materials were also composited with graphene to increase the performance of the CDI system, as reported by Zhang et al. [19]. The observed excellent desalination behaviour of the composite electrodes is due to their higher electrical conductivity and surface area. The above-mentioned CNT-based electrodes were made as two-dimensional (2D) electrodes. Hence, it is necessary to develop CNT-based three-dimensional (3D) electrodes for better effectiveness in a CDI system because of their highly porous surface area and large amount of ion diffusion [20].

One important material that is used in 3D electrodes is reticulated vitreous carbon (RVC), which does not possess a significant surface area compared to several other carbons, like activated carbon (AC), activated carbon nanofibre (ACF) web [21], carbon nanotubes (CNT) [22], composite carbon nanotubes with carbon nanofibre (CNT/CNF) [23], reduced graphite oxide (rGO) [24], and carbon aerogels (CA) [25], which are used as 3D CDI electrode materials. Hence, RVC can be used as a 3D template for all other materials because of its excellent properties, such as the reduction in the resistance of solution flow through the electrode, the increase in the stability of composite electrodes towards high flow-rate pressure, the increase in the possibility of ions to reach all electrode surfaces in a short time for electrosorption, and the reduction in the time to release the ions from the electrode surface, compared to the other 3D electrodes.

It is clear from the above discussion that studies on the development of composite electrodes in CDI systems based on RVC and CNTs is very rare. Previously, we reported how the 3D poly(3,4-ethylenedioxythiophene/reticulated vitreous carbon PEDOT/RVC electrodes [26,27] can be used efficiently in CDI systems where RVC was used to support PEDOT. This motivated us to use RVC electrodes to build 3D nanoweb-functionalized SWCNT electrode structures for CDI systems by coating RVC with SWCNT using a dip-coating methodology.

The objective of the present study is to prepare RVC/SWCNT electrodes and use them in the CDI system for desalination. In this study, acid functionalization of single-wall carbon nanotubes (a-SWCNTs) was done and then coated on the RVC electrode, which was then used as a 3D template/support. Actually, the pristine CNTs have a tendency to form aggregates within the composite system. Hence, to avoid agglomeration and to obtain better dispersion, the CNTs were functionalized. This is because the functional group will keep the CNT particles apart from each other by preventing the formation of aggregates. It also helps to form the porous structure. Moreover, functionalization converts CNTs into hydrophilic materials that reduce the flow resistance of aqueous solutions through the electrodes. These electrodes were tested in a capacitive deionization system and the performance of these electrodes in the system was investigated under various working conditions, such as the flow rate and bias potential, which were optimized. Moreover, the effect of the feed stream directly through the electrode or between the electrodes, and the distance between the electrodes in the CDI system, was investigated. Furthermore, the interaction of ions with the electrodes was investigated using the electrosorption isotherms, such as Langmuir and Freundlich models, and the performance of the electrodes was evaluated by electrosorption dynamics. The basic and improved design of flow-through cells made by 3D printing was also considered. All of these are important for developing electrodes and to use effectively in purification technology.

2. Materials, Methods, and Experiments

2.1. Chemicals and Materials

SWCNTs (Hipco-CCNI/Lot#p1001) were supplied by Carbon Nanotechnologies, Inc. (Houston, TX, USA) and were used as received. The solvent, *N,N*-dimethylformamide (DMF) (AR grade), concentrated HNO_3 (70%), and NaCl (AR grade) were purchased from Sigma-Aldrich (Darmstadt,

Germany) and were used as received. The RVC (compressed 60 ppi (pores per inch)), procured from ERG Materials and Aerospace Engineering (Oakland, CA, USA), was cleaned before use. Membrane filters (0.2 µm GTTP) were supplied by Millipore (Sigma-Aldrich Darmstadt, Germany). Milli-Q water possessing a resistivity of 18.2 MΩ cm was used in all preparations.

2.2. Method to Functionalize Carbon Nanotubes

The functionalization of SWCNTs (20 mg dispersed in 40 mL 6 M HNO$_3$) was carried out by refluxing in a round-bottom flask fitted with a reflux condenser and a magnetic stirrer. The reflux was performed by placing the flask in an oil bath at 120 °C for 6 h. The product was then filtered and washed with water until neutralization, and subsequently washed with methanol (10 mL) and DMF (10 mL), and then dried in an oven for 48 h at 105 °C. The acid-treated SWCNTs were designated as a-SWCNTs.

2.3. Pre-Treatment of the RVC Electrodes

Before treatment, all the RVC electrodes were cut into pieces of equal dimension using an RVC block. The dimension of each piece was 4 cm × 1.8 cm × 0.3 cm (length × width × thickness) = 2.16 cm^3. Initially, the electrodes were cleaned to remove impurities from their surfaces by soaking in 2 M HNO$_3$ for 24 h and then the acid was neutralized by washing several times with distilled water. Then the electrodes were soaked further in methanol for 2 h to remove any other type of organic impurities. The electrodes were then dried by flowing nitrogen and heating overnight at 110 °C. The weight of the electrodes was noted accordingly.

2.4. Method of a-SWCNT Dip Coating on RVC Electrodes

Initially, for dip coating, all the treated RVC electrodes were immersed into the 0.2% *w/v* a-SWCNT solution. The immersion was done slowly to allow the escape of air bubbles and prevent the formation of air pockets. After immersion, within a few seconds, the whole part of the RVC electrode was filled with solution due to capillary action. The electrodes were removed and dried initially at room temperature overnight, then at 100 °C for 2 h in an oven, and finally at 50 °C for 2 h using a vacuum oven to remove all organic solvents from the micropores of the electrodes. To achieve maximum loading of a-SWCNT on RVC, the process was repeated 2, 5, 10, and 20 times. The quantity of a-SWCNT loading on the RVC substrate was determined by weighing the electrode before and after dip coating. The loadings of a-SWCNT onto the RVC substrate were 6 mg (3.63 wt%), 23 mg (12.50 wt%), 34 mg (17.43 wt%), and 50 mg (23.85 wt%), when the process was repeated 2, 5, 10, and 20 times, respectively. The process of dip coating is schematically shown in Figure S1 (see Supplementary Section S1).

2.5. Electrochemical and Microscopic Characterisation

Cyclic voltammetry (CV) was used to determine the capacitance of the electrodes. The composite electrode, a-SWCNT/RVC, acted as a working electrode (WE) in 1 M aqueous NaCl solution. The scanning was done using a three-electrode system within the voltage range −0.2 to 1.0 V and a scan rate of 5–200 mV/s. In this measurement, RVC was used as a counter electrode (CE) and Ag/AgCl (3 M NaCl) as a reference electrode (RE). The electrical contact between the WE and the CE was made by the use of Pt wire. The procedure for measuring the amount of ion removal from the NaCl aqueous solution and on how to construct a CDI cell has been described in Supplementary Sections S2 and S3 along with Figures S2 and S3, respectively. The morphology of the a-SWCNT-coated RVC electrodes was analyzed by using a field emission scanning electron microscope (FESEM, ZEISS Sigma, Hamburg, Germany) at a specific voltage of 0.5 kV.

2.6. Different Working Conditions for Measuring Ion Removal Efficiency

The performance of ion removal of NaCl onto the sites of a-SWCNT is greatly affected by different working conditions, like flow rate, applied electrical voltage, loading level of a-SWCNT, and cell configuration. A systematic investigation of these different conditions on the efficiency of ion removal has been carried out in this section. These experiments were carried out with 60 mL NaCl solution. The concentration of NaCl was 75 mg/L whose measured electrical conductivity was 143.00 µS/cm. In this measurement, RVC was used as a counter electrode and the Ag/AgCl electrode as a reference electrode. The temperature was kept constant at 293 K during the experiment.

2.7. CDI Cell Configuration and Its Salient Features

While designing a CDI system, the fact that the feed stream to be desalinated should flow primarily between the two porous electrodes is considered [19,28–30]. The separation between the electrodes would be such that it acts as a flow channel and also prevent the electrical short circuit between the electrodes. Flow between (FB) electrodes in CDI systems require long desalination times [31]. Therefore, an attempt was made to change the location of the electrodes so that there is direct flow of the feed stream through the electrode pores to reduce the desalination time. This flow system, now called the flow-through (FT) electrode cell, often uses low hydraulic resistance and high surface area porous electrodes that consist of micropores and nanopores [31]. According to Suss et al. [31], when carbon aerogel is used in an FT cell, the mean sorption rate increases 4–10 times higher than that typically achieved with an FB cell. This is because the FT cell allows ion transport from the microscale pore bulk to the nanoscale pores, which maximizes the surface area [31]. Figure 1a,b represents the configuration of a typical FB charging electrode through a porous separator element and an FT capacitive desalination cell that allows the feed solution to flow directly through the electrode pores, respectively. The electrosorption performance of both cells was compared.

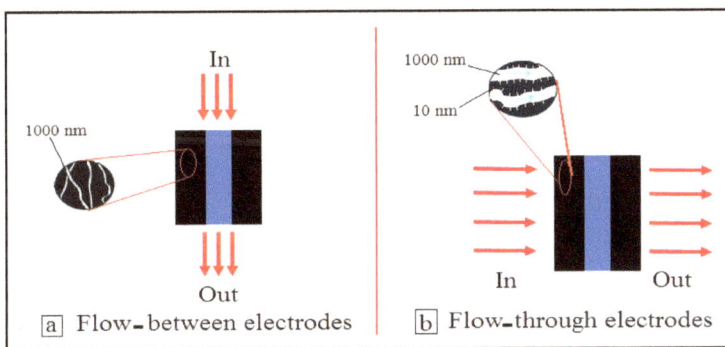

Figure 1. Schematic diagram of flow-between and flow-through electrodes in CDI cells.

2.8. Working Conditions for Investigating the CDI Cell Configuration on Ion Removal Efficiency

In all the experiments 60 mL of the 75 mg/L NaCl solution was used and it was re-circulated at 50 mL/min as a feed solution in the CDI system under an applied voltage of 1.5 V. An a-SWCNT 50 mg coated RVC electrode (23.58 wt%) was used as the working electrode. Additionally, the distance between the electrodes in both configurations was 5 mm.

2.9. Designing a New Flow-Through Cell

A new CDI cell was designed based on the flow-through electrode configuration aiming for obtaining a faster desorption cycle. Figure 2 shows the schematic of a new design made in our laboratory. It was built on a Connex 350 3D printer by Objet. This time the flow-through cell was produced using the MED610 material printed over a period of 3 h. This cell was rectangular shaped outside and the dimensions were 50 mm × 70 mm × 28 mm in height, length, and width, respectively (Figure 2a). Each side had one 4 mm diameter hole in the middle that served as the inlet and outlet ports for re-circulating the fluid flow by pumping. When entering the cell, the solution was passed through a flow distributor chamber 40 mm × 20 mm × 10 mm in height, length, and width, respectively, that had 45 (0.35 mm diameter) holes in the exit side that helped to direct flow onto the whole electrode surface (Figure 2b). A similar chamber was constructed for the outlet end. The cell was designed inside to fit a series of electrodes and maintain separation between them. The dimensions of each location and each separating space were 50 mm × 22 mm × 5 mm and 50 mm × 20 mm × 5 mm in height, length and width, respectively. This flow-through cell can hold up to five electrodes at any given time. The total volume of the solution was increased in this cell to 70 mL while the old cell was 60 mL. It is important to mention that the volume of the solution is a parameter in the electrosorption capacity equations (see Equations (S2)–(S4), Supplementary Section S4).

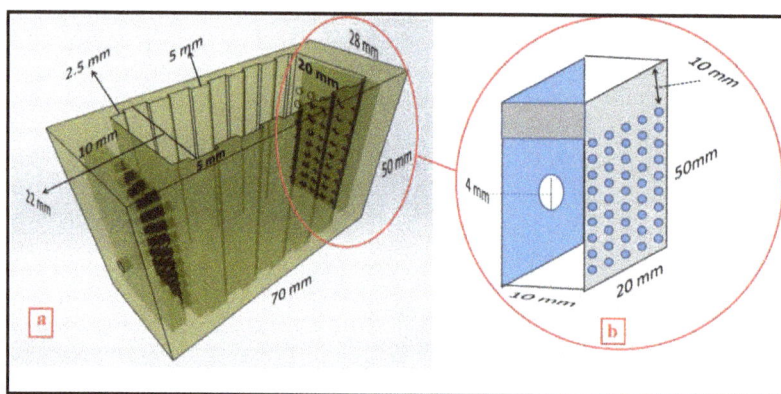

Figure 2. Schematic of a flow-through cell (**a**) and cross-section of flow-distributor chamber (**b**).

3. Results and Discussion

3.1. Scanning Electron Microscopy (SEM) and Cyclic Voltametry (CV)

The details of SEM and CV studies for RVC and CNT/RVC composites are presented in Supplementary Sections S5–S9. SEM was used to observe the morphology and calculate the pore size of RVC and CNT/RVC composite electrodes (shown in Figures S4 and S5). It was shown that the average pore size was 350, 700, and 900 μm for RVC electrodes having porosities 60, 45, and 30 ppi, respectively. The RVC electrode, containing a porosity of 60 ppi, was selected as the optimum electrode for loading the a-SWCNT because it showed the highest specific capacitance value and surface area per volume of the electrode. It was revealed that the pore was of the macroscale, through which the ions can diffuse easily. The specific capacitance was calculated through cyclic voltammetry (shown in Figures S6 and S7, and Table S1). It was increased by a factor of 280, 510, and 655 for 3.63, 12.50, and 17.43 wt% a-SWCNT/RVC electrodes, respectively, compared to a bare RVC electrode. The specific capacitance value was observed to gradually decrease with the increase in scan rate when tested for the 3.63 wt% a-SWCNT/RVC electrode. Moreover, the specific capacitance value, when measured per geometric volume, was seen to increase with the increase in the amount of a-SWCNT coating on RVC, indicating the increment in the coated surface area of a-SWCNT on the RVC electrode.

3.2. Effect of Applied Voltage on Ion Removal Efficiency

In this study, the 3.63 wt% loaded a-SWCNT/RVC composite acted as a working electrode in the CDI system. The experiment was carried out within a voltage range of 0.9–1.5 V. The choice of this voltage range is based on some previous studies so that we can compare our results with the results reported in the literature [21–25]. The comparison has been made in the last section of this article. The variation in electrical conductivity of the NaCl solution with respect to the mentioned voltage range and operation time is shown in Figure 3a, whereas Figure 3b shows the variation of electrosorption capacity (calculated as explained in Supplementary Section S4) against the electrical voltage. The ion removal characteristics were affected by various applied voltages. With the applied electric field, there is a dramatic drop of salt concentration because the ions present in the system feel attraction by oppositely-charged electrodes [32]. As the applied voltage increased in the range of 0.9–1.5 V, the ion removal amount increased and the electrosorption capacity gradually rose from 3.15 mg/g to 8.92 mg/g. This clearly indicates that the electrosorption capacity increases with the increase in applied voltage, and this can be attributed to the strong Coulombic interaction originating between the electrodes and charged Na^+ and Cl^- ions [24,33]. Electrolysis of water was also not observed at and above 1.2 V, as evident from the non-visibility/non-appearance of any gas bubbles from the solution, which can account for the existing resistance within the whole circuit [34,35]. It is observed from the plot that the electrical conductivity of the NaCl solution decreases approximately to 141.34 µS/cm, 141.61 µS/cm, 142.43 µS/cm, and 142.67 µS/cm at an applied voltage of 1.5 V, 1.3 V, 1.1 V, and 0.9 V, respectively. The plot also suggests that an efficient CDI process occurs at 1.5 V due to the enhancement of electrostatic force, whereas poor performance is observed at 1.1 V. When there is no variation in conductivity, it indicates that the ions were adsorbed maximally [32]. The electrosorption phenomenon exhibits its reversibility, as is evident after switching off the electric field, which results in the returning of conductivity to its initial state. Thus, it can be inferred that electrosorption processes in CDI cells and applied voltage are inter-related phenomena.

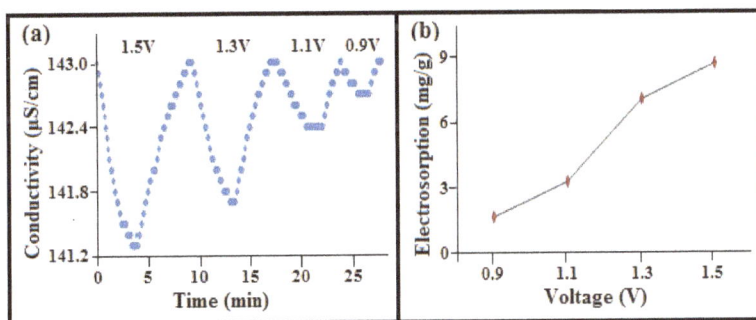

Figure 3. (a) Conductivity variations of NaCl solution in the adsorption and release time for 3.52 wt% a-SWCNT-coated RVC electrodes at a flow-rate of 25 mL/min with various applied voltages and operating times; and (b) electrosorption as a function of applied voltage.

3.3. Role of the Flow-Rate on Electrosorption

The effect of increasing the flow-rate on the electrode's ability to adsorb ions was investigated over a flow rate of 25–75 mL/min through a CDI cell as depicted in Figure 4. It is obvious that the efficiency of the removal of NaCl ions using 3.63 wt% a-SWCNT and 12.50 wt% a-SWCNT coating RVC electrodes decreases after increasing the flow rate above 50 mL/min as shown in Figure 4a,b. The maximum decrease in conductivity of the NaCl from an initial value of 143 µS/cm, using 3.63 wt% and 12.50 wt% a-SWCNT/RVC electrodes, is around 140.72 µS/cm and 140.14 µS/cm at a flow-rate of 25 and 50 mL/min, respectively. The conductivity at a flow rate of 75 mL/min was decreased to 141.33 and 140.82 µS/cm using 3.63 wt% and 12.50 wt% a-SWCNT/RVC electrodes, respectively. These results indicate that the flow rates above 50 mL/min would lead to lower electrosorption. This is because a high pump rate will exert a high pump force and if this force becomes greater than the electrosorption force then there will be decrement in the amount of electrosorption [28,34]. In addition, the conductivity characteristics related to flow rates were not significantly changed after the increment in flow rate from 25 to 50 mL/min. This is because the electrostatic force of the electrode reaches the equilibrium condition with the driving force of the flow rate [28,34]. Therefore, the amount of electrosorption did not significantly change in both cases. It is clear that when the loading level of a-SWCNT increased in the RVC electrodes to 17.43 wt% and 23.58 wt% (Figure 4c,d), the efficiency of the conductivity at the low flow rate (25 mL/min) was worse. The reason is that a low pump rate results in a co-ion effect that leads to depression of the electrosorption process [28,34]. Furthermore, the result shows that 50 mL/min is the optimal flow rate, as shown in Figure 4c,d.

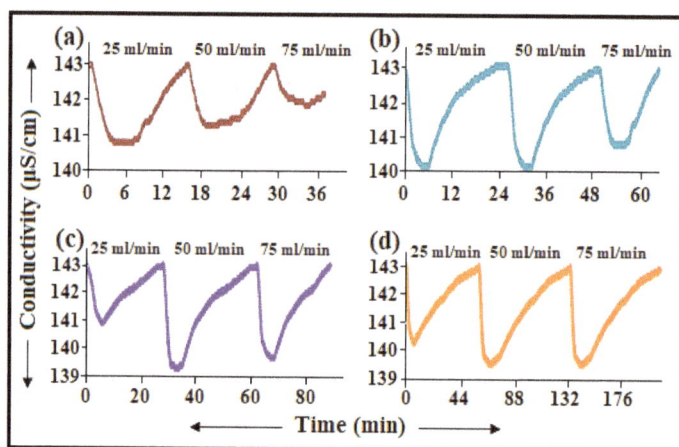

Figure 4. Variations in the conductivity of in the adsorption and release of NaCl solution at different applied flow rates and operating times when measured at 1.5 V: (**a**) 3.63 wt%; (**b**) 12.50 wt%; (**c**) 17.43 wt%; and (**d**) 23.58 wt% a-SWCNT-coated RVC electrodes.

Thus, it is evident that a voltage of 1.5 V and a flow rate of 50 mL/min are the optimum parameters exerting a positive effect on the ion removal performance of NaCl onto the sites of a-SWCNTs. Hence, for investigating the role of a-SWCNT loading on NaCl ion removal efficiency, these two parameters were taken as optimum conditions as discussed and reported in the later section of this article.

3.4. Effect of a-SWCNT Loading on Electrosorption

The effect of increasing a-SWCNT loading on a-SWCNT/RVC composite electrodes on the ion removal performance was investigated at the loading levels of 3.63, 12.50, 17.43, and 23.58 wt%, respectively, and is depicted in Figure 5a. We choose these arbitrary loading levels to investigate the loading levels of SWCNTs on the electrosorption capacity, as well as to check the effectiveness of our electrodes compared to some published literature [21–25]. This has been discussed in the last section of this article. The figure shows that the application of electrical voltage decreases the conductivity of all electrodes because ions were attracted by opposite charges on the electrodes [32]. Then the conductivity approaches a minimum value, which implies that the system is saturated. These adsorption processes, at first, took 6 min. It is clear that the drop in conductivity of the solution increases with the increase in the amount of a-SWCNTs on the RVC electrode. This is because of the increment in loading of a-SWCNTs, which results in the increase in interaction between the charged electrode surfaces and

Na$^+$ and Cl$^-$ ions. Notably, the highest drop in conductivity was 4.41 µS/cm using the electrode, which had 23.58 wt% a-SWCNT. For other electrodes, the conductivity decreased from 143.00 µS/cm to 141.34 µS/cm, 140.32 µS/cm, and 139.29 µS/cm with a-SWCNT loadings of 3.63 wt%, 12.50 wt%, and 17.43 wt%, respectively. Moreover, when the voltage of the CDI system is reduced to 0 V, the electrodes are found to be quickly regenerative. This indicates that the adsorbed ions on the electrodes are desorbed because of the disappeared electrostatic force.

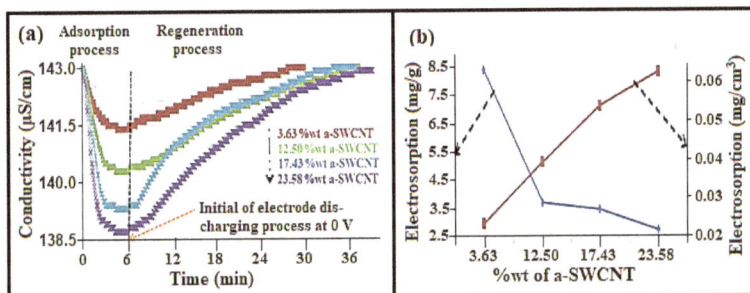

Figure 5. (**a**) Adsorption and release behaviour of various a-SWCNT-coated RVC electrodes at 1.5 volt, using 50 mL/min water flow-rate between electrode; and (**b**) the electrosorption capacity in terms of mass of a-SWCNTs and in terms of the geometric volume of the composite electrode at various loadings of a-SWCNTs on the RVC electrode using the CDI system.

It had taken almost 30 min to discharge all the ions from the electrode and to reach the conductivity at its initial state. Hence, it can be said that the CDI system made using the a-SWCNT/RVC composite can be effectively used for desalination technology.

The electrosorption removal of NaCl by our CDI system was measured from the data in Figure 5a. Hence, a calibration was made to find a relationship between conductivity (µS/cm) and concentration (mg/L) before conducting the experiment (see Equation (S1), Supplementary Section). The electrosorption behaviour of a-SWCNT-coated RVC electrodes is presented in Figure 5b, which shows that the electrosorption decreases with the increase in mass of a-SWCNT. It is clear that when the mass of a-SWCNT in 3.63 wt% a-SWCNT/RVC and 23.58 wt% a-SWCNT/RVC electrodes was 6 mg and 50 mg, the electrosorption was 8.39 mg/g and 2.77 mg/g, respectively. On the other hand, the value of electrosorption increases with the increase in the amount of a-SWCNT when the electrosorption of the electrode is considered in terms of the geometric volume or geometric area of the a-SWCNT coated on RVC. It is estimated that when the mass of a-SWCNT in 3.63 wt% a-SWCNT/RVC and 23.58 wt% a-SWCNT/RVC electrodes was 6 mg and 50 mg, the electrosorption was 0.003 mg/cm^2 or 0.02 mg/cm^3, and 0.008 mg/cm^2 or 0.06 mg/cm^3, respectively. Additionally, it is clear that the electrosorption became more stable above an electrode loading of 12.50 wt% a-SWCNT. Table 1 shows more details of the electrosorption in terms of mass, area, and volume for each a-SWCNT-coated RVC electrode. Hence, it can be inferred that the electrosorption performance of the 3.63 wt% a-SWCNT/RVC electrode is the best in terms of mg/g of ion removal, whereas the electrode with 23.58 wt% a-SWCNT loading exhibits the best ion removal efficiency in terms of either geometric area or geometric volume of the electrode. In other words, it can be said that the removal of ions is more dependent on the electrode size, which is a great advantage when one considers the size of the electrode to design a CDI system. Having designed a simple CDI system and investigated its electrosorption performance, attention was then turned towards refining the CDI system with an improved cell design. This work is now reported in the next section.

Table 1. Electrosorption capacity of various a-SWCNT-coated RVC electrodes.

Sample	a-SWCNT in Sample (wt%)	Electrosorption Capacity		
		mg/g of a-SWCNT	mg/cm^3 of Electrode	mg/cm^2 of Electrode
1	3.63	8.39	0.02	2.8×10^{-3}
2	12.50	3.69	0.03	4.8×10^{-3}
3	17.43	3.42	0.05	6.5×10^{-3}
4	23.74	2.77	0.06	7.6×10^{-3}

3.5. Effect of CDI Cell Configuration on the Removal Efficiency of Ions

Figure 6 shows the effect of feed stream for the flow-through (FT) electrode and the flow-between (FB) electrode with respect to the desalination time for the 23.58 wt% a-SWCNT-coated RVC electrode measured at a flow rate of 50 mL/min and a voltage of 1.5 V. It is observed that, to complete one desalination cycle, it takes 39 min and 18 min when the flow-between electrode and flow-through electrode configurations are used, respectively. In fact, the time required for desalination decreased more than two times when the flow-through electrode configuration was applied. In both systems, the adsorption process took 6 min, and the rest of the time was required to regenerate the electrodes. This means that the release of salt from the electrode required 12 min and 33 min for flow-through electrodes and flow-between electrodes, respectively. This improvement in the salt removal and regeneration process is due to the facilitation in ion transport at the electrode solution interface and faster electron transport within the electrode that resulted in the three times faster desalination cycle when the solution was flowed directly through the electrodes. It is very important to draw attention to the shape of the conductivity curve, although the amount of conductivity decrease was the same in both systems. The conductivity after 6 min in both systems was 138.59 µS/cm. This can be attributed to the flowing force of ion transport to the electrode interface, which was the same in both systems.

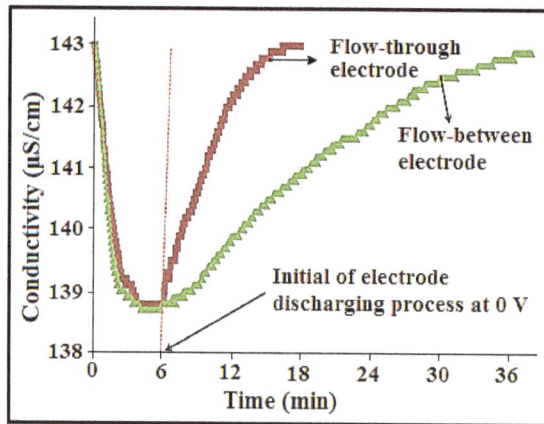

Figure 6. Comparison of NaCl adsorption and the release behaviour of the 23.58 wt% a-SWCNT-coated RVC electrode when configured for flow-between and flow-through systems in a CDI cell measured at a flow rate of 50 mL/min, 75 ppm NaCl, and 1.5 V.

3.6. Effect of the Flow Rate and Voltage on Ion Removal Efficiency

In the new cell system, the total volume of solution was increased to 70 mL. The concentration and the electrical voltage were similar to the previous experiments (75 mg/L NaCl solute ion and 1.5 V). The flow rate was investigated to study the NaCl removal using the best performing electrode in the old system, which was the 23.58 wt% a-SWCNT-coated RVC electrode. The distance between the electrodes was 5 mm. According to the old system, the best performance for water purification was at a flow rate of 50 mL/min. This led to the investigation of electrode performance at flow rates below and above 50 mL/min, as shown in Figure 7. It is obvious from the results that the best conductivity decrease was achieved when the flow rate was at 50 mL/min. When the flow rate was above or below 50 mL/min, the conductivity decrease was less, leading to lower electrosorption capacity. This depression in electrosorption capacity below a flow rate of 50 mL/min is because of the co-ion effect resulting from a low pump rate, while above 50 mL/min, the pump force is higher than the electrosorption force at a high pump rate, thereby decreasing the amount of electrosorption [34]. Thus, the optimum flow rate for the CDI process was found to be 50 mL/min.

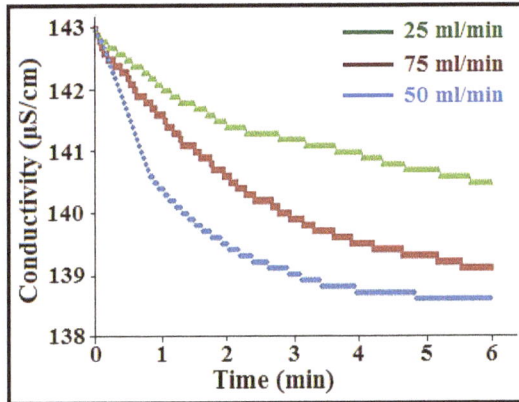

Figure 7. Conductivity variations of the NaCl solution for the 23.58 wt% a-SWCNT-coated RVC electrode as a function of operation time with respect to various flow rates (25 mL/min, 50 mL/min, and 75 mL/min) measured at 1.5 V.

In the later sections, all experiments have been carried out with 70 mL of the 75 mg/L NaCl solution at a flow rate of 50 mL/min through the CDI system at an electrical voltage of 1.5 V.

3.7. Effect of the Distance between the Electrodes on the Efficiency of Electrosorption

It is known that increasing the space between a pair of electrodes results in an increase in electrical resistance between the electrodes that, in turn, leads to a decrease in the electrical current [36,37]. Therefore, the next phase of the study was directed towards the effect of current density and space between the electrodes on the efficiency of electrosorption. The space between the electrodes was varied from 5 mm to 15 mm and then to 25 mm, respectively, the counter electrode was the RVC electrode, and the working electrode was the 23.58 wt% a-SWCNT-coated RVC electrode. Figure 8 shows the current and conductivity behaviour of the 23.58 wt% a-SWCNT/RVC composite electrodes at various gaps between the electrodes. As expected, the current density decreased with an increase in the gap between the electrodes. When the distance between the electrodes was 5, 15, and 25 mm, the current at the start was found to be 4.56, 2.20, and 1.11 mA, respectively (Table 2). Table 2 also gives more details about the start current, current stability, charge of the electrode, the time of highest conductivity, and the energy output for each distance between the electrodes. Furthermore, there is a variation in the output energy of the CDI system when the distance between the electrodes is increased. It is clear that if the space increases, the energy output increases. The energy output at 5, 15, and 25 mm was 0.67, 0.91, and 1.42 J/C, respectively. The energy is calculated according to Equation (1) [38]:

$$E = Q \times V \tag{1}$$

where E is the energy (J), Q is the charge (C), and V is the constant voltage (V).

Table 2. The current of the CDI system for different inter-electrode distances using 1.5 V.

Distance between Electrodes (mm)	Time Required for Adsorption (min)	Initial Current (mA)	Stable Current (mA)	Charge (C)	Energy (J)
5	6	4.56	1.02	0.45	0.67
15	25	2.20	0.27	0.61	0.91
25	60	1.11	0.16	0.95	1.42

It is very interesting to note that the performance of ion removal is affected when the space between the electrodes is varied. With the increase in distance between the electrodes, the amount of ions removed was not affected in the range of distances from 5 mm to 15 mm, but was less affected when the gap between the electrodes was 25 mm. However, there is an increment in the adsorption time as the distance increases in all cases. One deionisation cycle took 7 min at a distance of 5 mm between the electrodes, but when the distance was increased to 15 mm the ion adsorption/desorption cycle took 55 min, as shown in Figure 8. As mentioned above, the best design should have a small space between the electrodes because this will reduce the time of water purification, afford efficient energy output, decrease the electrical resistance between the electrodes, and increase the electrical current.

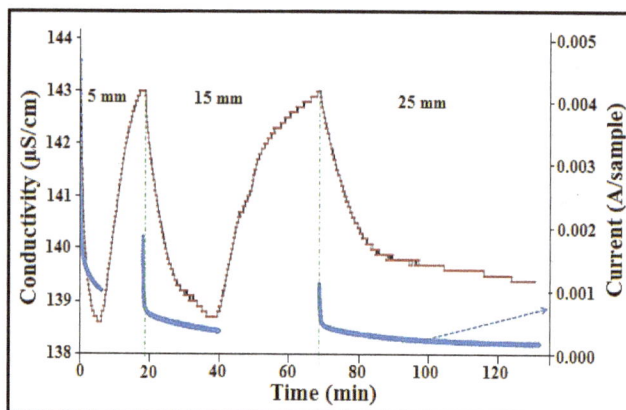

Figure 8. The effect of distance between the electrodes (5 mm, 15 mm, and 25 mm) on the current and conductivity behaviour of the 23.58 wt% a-SWCNT-coated RVC electrode measured at a voltage of 1.5 V and a flow rate of 50 mL/min.

3.8. Adsorption/Desorption Performance of a-SWCNT Loaded RVC Electrodes

Figure 9a shows the CDI process of adsorption and desorption performance at all loading levels of a-SWCNT in RVC electrodes experimented at an optimized electrode distance, electrical voltage, and flow rate. It is clear that the electrosorption behaviours of the electrodes measured using the new cell followed the same trend like the old cell, as mentioned earlier; that is, the drop in conductivity increases with the increase in the amount of a-SWCNT on the electrodes. It is noticed that for the 23.58 wt% a-SWCNT-coated RVC electrode, there is a decrease of about 4.41 µS/cm in electrical conductivity, indicating a significant achievement in electrosorption performance. Initially, for all electrodes, the conductivity gradually decreases up to a minimum value, which indicates that saturation was reached. These adsorption processes took 6 min. During discharging of the CDI system at 0 V, the conductivity returned nearly to its initial state (143 µS/cm). This is an indication to the release of ions from their electrical double-layer region back into the solution because of the disappearance of electrostatic forces, as have been discussed earlier. It took 18 min for complete discharge of the CDI cells for the electrode with the highest amount of a-SWCNT in the sample (23.58 wt%). This result is in agreement with the earlier reported results in Figure 5a.

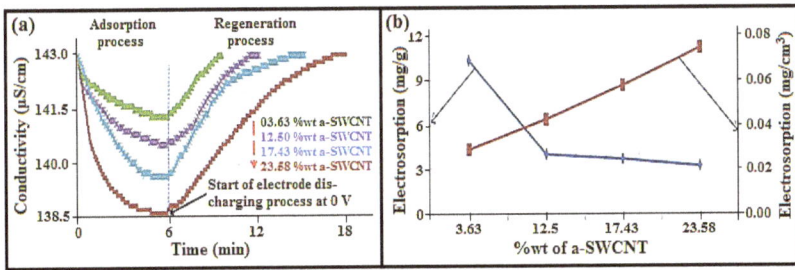

Figure 9. (**a**) Adsorption and release behaviour; and (**b**) the electrosorption capacity in terms of the mass of a-SWCNT and the geometric volume of the electrode at various a-SWCNT-coated RVC electrode coatings of a-SWCNT (wt%): 3.63%, 12.50%, 17.43%, and 23.58%.

Figure 9b shows the electrosorption behaviour of different a-SWCNT/RVC composite electrodes in terms of the mass of a-SWCNT and the volume of the electrode. It is clear that the electrosorption decreases with the increase in the weight of a-SWCNT loading. Clearly, when the sample had 3.63 wt% a-SWCNT, the electrosorption capacity was 10.40 mg/g and when the sample had 23.58 wt% a-SWCNT, the electrosorption capacity was 3.23 mg/g. On the other hand, if the electrosorption of electrodes was considered in terms of the geometric area or geometric volume, the electrosorption increased with the increase in the amount of a-SWCNT. For example, when the sample had 3.63 wt% a-SWCNT, the electrosorption was 0.03 mg/cm^3 or 0.003 mg/cm^2, and when the sample had 23.58 wt% a-SWCNT, the electrosorption was 0.08 mg/cm^3 or 0.009 mg/cm^2 (Table 3). Table 3 also gives more details about electrosorption in terms of mass, geometric area, and geometric volume for each a-SWCNT/RVC composite electrode.

Table 3. Electrosorption capacity of different a-SWCNT/RVC composite electrodes.

Sample	a-SWCNT in Sample (wt%)	Electrosorption Capacity		
		mg/g of a-SWCNT	mg/cm^3 of Electrode	mg/cm^2 of Electrode
1	3.63	10.40	0.03	3.5×10^{-3}
2	12.50	3.99	0.05	5.1×10^{-3}
3	17.43	3.67	0.06	7.2×10^{-3}
4	23.58	3.23	0.08	9.4×10^{-3}

3.9. Comparison between the New Cell (Flow Feed-Through Electrode) and the Old Cell (Flow Feed-Between Electrode)

This section compares the results of electrosorption capacity and the time of one desalination cycle, obtained with various a-SWCNT-coated RVC electrodes, for two different flow cells (flow feed-through electrodes and flow feed-between electrodes), as shown in Table 4. During the experiment, the concentration of the NaCl feed solution was 75 mg/L. Table 4 shows that the electrosorption capacity of electrodes has increased when measured using the new cell. The increment in elecrosorption capacity is from 8.39 mg/g to 10.40 mg/g, which is 23.96% (see example calculation below) if we consider the 3.63 wt% a-SWCNT/RVC composite electrode. In fact, there is approximately a three times decrement in the desalination cycle time, for example, from 30 min to 10 min, from 36 min to 12 min, from 37 min to 15 min, and from 39 min to 18 min when measured using the new cell for 3.63 wt% a-SWCNT, 12.50 wt% a-SWCNT, 17.43 wt% a-SWCNT, and 23.58 wt% a-SWCNT-coated RVC electrodes, respectively.

Table 4. Electrosorption capacities and the time of one desalination cycle of various a-SWCNT-coated RVC electrodes obtained by comparing the performance between flow feed-between (FB) electrodes and flow feed-through (FT) electrodes with the NaCl feed solution (75 mg/L).

a-SWCNT in the RVC Electrode (wt%)	Flow Direction	Electrosorption (mg/g of a-SWCNT)	# Enhancement in Electrosorption (%)	Time of One Desalination Cycle (min)
3.63	FT	10.40	23.96	10
	FB	8.39	-	30
12.50	FT	3.99	8.13	12
	FB	3.69	-	36
17.43	FT	3.67	7.31	15
	FB	3.42	-	37
23.58	FT	3.23	16.61	18
	FB	2.77	-	39

An example calculation is given in Supplementary Section S10.

3.10. CDI Cycling Stability

Regeneration of the a-SWCNT/RVC electrode is very important because of its practical applicability in CDI systems. The 23.58 wt% a-SWCNT electrode was used in this study because, previously, it exhibited the highest electrosorption capacity when measured per geometric volume using the flow feed-through cell. Figure 10 shows the electrosorption/regeneration cycles of the 23.58 wt% a-SWCNT-coated RVC electrode, which was performed repeatedly several times by charging and regeneration cycles. In the absence of the oxidation–reduction process in electrosorption, the consumption of current is basically to charge the electrode by electro-adsorbing the ions from the bulk of the solution [39]. It is seen from the figure that, initially, the solution conductivity when measured at 1.5 V sharply decreases because of the migration of ions onto the surfaces of oppositely-charged electrodes, and then there is a gradual decrement until the complete formation of an electrical double layer at the electrode/electrolyte interface [40]. The electrode is regenerated by depolarizing the system at 0.0 V. Herein, we have conducted four cycles of repeated electrosorption and desorption. It is observed that each cycle takes 17 min, which means 6 min to adsorb ions and 11 min to release ions, indicating that each cycle is going to complete in a short time period. Moreover, the stability of cycles is also high as there is no decay in electrosorption capacity. This reveals that the process is reversible and the composite electrode can be reused several times by controlling the amount of electro-adsorbed ions by manipulating the formed electrical double layer at the electrode/electrolyte interface.

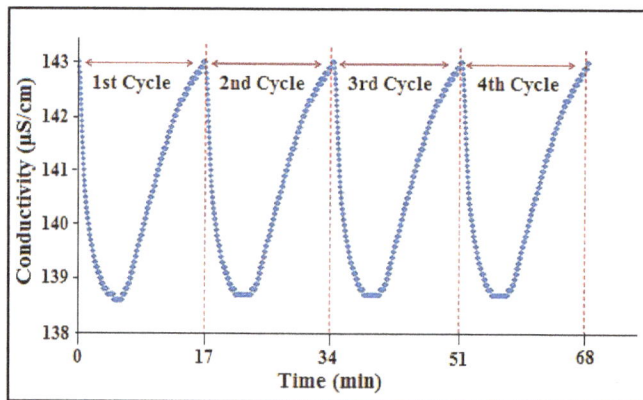

Figure 10. Multiple electrosorption–desorption cycles of 75 ppm NaCl solution for the 23.58 wt% a-SWCNT-coated RVC electrode. Polarization and depolarization were performed at 1.5 V and 0.0 V, respectively.

3.11. Electrosorption Dynamics

The different electrosorption dynamics models for NaCl adsorption onto the 23.58 wt% a-SWCNT/RVC composite electrode measured at 1.5 V and a flow rate of 50 mL/min using a new cell are presented in Figure 11. The studied electrosorption dynamics models are pseudo-first-order, pseudo-second-order, and intra-particle diffusion dynamic models. The electrosorption of NaCl onto the electrode was very rapid within the first minute, then it became dynamic adsorption and after 3 min the electrode gradually approached saturation, as shown in Figure 11a. It takes 6 min to reach the electrosorption at equilibrium. This may be because of the high diffusion rate of ions onto the a-SWCNT particle surfaces. When the Weber and Morris model (see Equation (S9) in Supplementary Section S11) [41] is used to study the intra-particle diffusion of ions into the electrode, it is clear that the plot of q_t versus $t^{0.5}$ shows three multi-linear regions, as shown in Figure 11b. This behaviour is quite common in an electrosorption process [28,34,42–44] and it indicates that more than one step is effective in the electrosorption process. When the voltage was applied, the first-stage adsorption occurred rapidly because the surface diffusion was transported from the bulk solution to the external surface of the sorbent [45]. In the second stage of adsorption, the adsorption occurs gradually in a linear manner. In this stage, the diffused intra-molecular sorbate molecules move more into the interior part of the sorbent particles. The third linear stage indicates that a low adsorption stage has occurred where intra-particle diffusion starts on the interior sites of the sorbent [45]. The intra-particle diffusion rate constant, k_{id}, calculated from the linear slope from stage 2 and the coefficient of correlation are presented in Table 5 [46]. These two parameters are 1.66 mg/gmin$^{0.5}$ and 0.981, respectively.

Table 5. Different parameter values obtained using the pseudo-first-order model, pseudo-second-order model, and intraparticle diffusion.

Pseudo-First-Order			Pseudo-Second-Order				Intraparticle Diffusion	
q_e (mg/g)	K_1 (min^{-1})	R^2	q_e (mg/g)	K_2 (g/mg min)	R^2	h (mg/g min)	K_{id} (mg/g min$^{0.5}$)	R^2
3.19	0.82	0.994	4.72	0.123	0.950	2.74	1.66	0.981

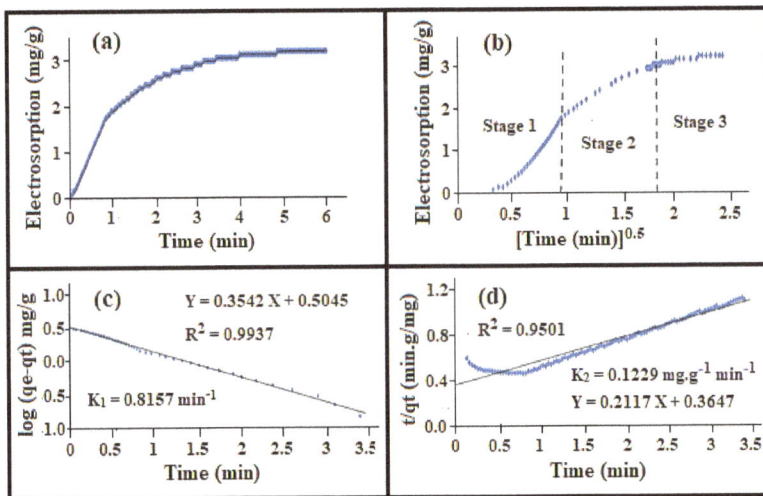

Figure 11. The (**a**) electrosorption; (**b**) intraparticle diffusion; (**c**) pseudo-first-order adsorption kinetics; and (**d**) pseudo-second-order adsorption kinetics of NaCl onto the 23.58 wt% a-SWCNT-coated RVC electrode at a flow rate of 50 mL/min and a voltage of 1.5 V.

For evaluating the kinetics of the electrosorption process, the pseudo-first-order and pseudo-second-order processes were studied within the first 3 min, as shown in Figure 11c,d, respectively. The data of log $(q_e - q_t)$ were plotted against t where the slope gives the value of first-order rate constant k_1 and the intercept represents the equilibrium adsorption density q_e, as shown in Figure 11c. It is seen that the linear fit has a relatively high correlation coefficient (R^2 value), suggesting that the application of Equation (S10) (see Supplementary Section S12) is appropriate as the plotted experimental data are in good agreement with their linear fit. The extracted parameters are reported in Table 5, which shows that the value of R^2 and k_1 for the pseudo-first-order kinetic model are 0.994 and 0.82 min^{-1}, respectively. The obtained value for q_e (3.19 mg/g) is also reasonable according to this kinetic model.

To determine the adsorption parameters of the pseudo-second-order process, namely q_e and k_2 as mentioned in Equation (S11) (see Supplementary Section S13), t/q_t was plotted against t as shown in Figure 11d. The figure shows that there is poor agreement between the experimental data and the theoretical values obtained using this model. The parameter values obtained from the second-order kinetic model are also tabulated in Table 5 and compared with the first-order kinetic model. The values of R^2 and k_2 for the second-order kinetic model obtained were 0.950 and 0.12 g/mg min, respectively, which revealed that the R^2 value in the second-order kinetic model is poor compared to the first-order kinetic model (0.996). The calculated q_e value is also not in good agreement with the experimental data (it is 4.72 mg/g). The above results suggest that the NaCl adsorption process follows the first-order kinetic model and confirms that the ions are not adsorbed onto the surfaces of a-SWCNTs via chemical interaction [45]. Many authors have reported a similar type of result in the past where NaCl ions were adsorbed from the aqueous solution by different adsorbents [7,24,28,34,42–44].

3.12. Electrosorption Isotherms

Electrosorption isotherms are essential to estimate the electrosorption behaviour of carbon electrodes. The electrosorption isotherm of NaCl was experimented onto the a-SWCNT (23.58 wt%)-coated RVC electrode at a voltage of 1.5 V, a flow-rate of 50 mL/min, and a temperature of 298 K using the CDI unit cell at different NaCl solution concentrations, starting from 25 mg/L to 500 mg/L, as shown in Figure 12. The electrosorption of a-SWCNT was affected at different initial solution concentrations. The figure shows that the increase in the solution feed concentration increases the removal of NaCl, which means the increase in electrosorption behaviour. This is because of the enhancement of the mass transfer rate of ions within the micropores and the subsequent reduction in the overlapping effect due to the increment in the concentration of solution [43,47,48]. It is noticed that at a feed solution concentration of 500 mg/L, the electrosorption capacity of the 23.58 wt% SWCNT-coated RVC electrode is 8.89 mg/g.

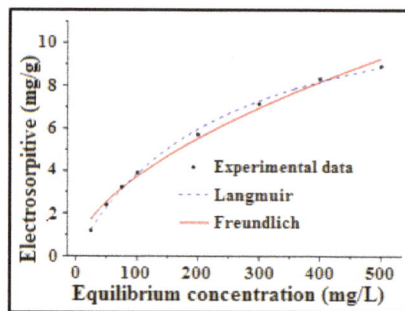

Figure 12. The electrosorption isotherm for the 23.58 wt% a-SWCNT-coated RVC electrode at a flow rate of 50 mL/min NaCl and a voltage of 1.5 V.

The experimental results were fitted to Langmuir and Freundlich isotherms (Equations (S12) and (S13), respectively; see Supplementary Section S14) for evaluating the electrosorption behaviour of Na^+ and Cl^{-1} onto the electrodes. Table 6 shows the value of the extracted parameters, such as the coefficients of correlation R^2, the Langmuir constant K_L related to binding energy, and the Freundlich constant K_F related to the adsorption capacity of adsorbent for NaCl electrosorption using the 23.58 wt% a-SWCNT-coated RVC electrode as an electrosorption electrode. It is found that the plots of electrosorption obtained using both isotherm models are in good agreement with the experimental plot for this electrode, which is also evident from their R^2 values, which is 0.997 for the Langmuir model and 0.989 for the Freundlich model. Thus, the monolayer adsorption can be suggested as the primary mechanism for this electrosorption adsorption process [7,28]. The K_L and K_F values of the a-SWCNT electrode were 0.01 and 0.28, respectively. Normally, for high adsorption, the value of n, which indicates the tendency of the adsorbate to be adsorbed should be between 1 and 10 [34]. The value of n in this electrode was 1.77, indicating that the a-SWCNT electrode exhibits a high potential electrosorption capability. One can obtain a very thin layer of adsorbate for such a system where the adsorbed amount will be a fraction of the monolayer capacity. Hence, the electrosorption of the a-SWCNT/RVC electrode in the CDI system follows the monolayer adsorption [42]. Moreover, for estimating the maximum electrosorption amount of a-SWCNT, the Langmuir isotherm model was selected. The parameter q_m in this model accounts for the maximum adsorption capacity, which suggested that q_m was improved as the bias concentration increase. The equilibrium electrosorption capacity of this electrode at the above mentioned experimental conditions was 13.08 mg/g.

Table 6. Determined parameters of regression coefficients R^2, K_L, and K_F of Langmuir and Freundlich isotherms for NaCl electrosorption by using the a-SWCNT (23.58 wt%)-coated RVC electrode as an electrosorption electrode.

Isotherm	Parameter	Value
Langmuir	q_m	13.08
	K_L	0.01
	R^2	0.997
Freundlich	K_F	0.28
	n	1.77
	R^2	0.989

3.13. Comparison of the Present Work with the Previously-Published Research Studies

In this section, the electrosorption capacities of a-SWCNT-coated RVC electrodes were compared with several other carbon materials, such as activated carbon (AC), activated carbon nanofibre (ACF) web [21], carbon nanotubes (CNT) [22], composite carbon nanotubes with carbon nanofibre (CNT/CNF) [23], reduced graphite oxide (rGO) [24], and carbon aerogels (CA) [25], which are used as CDI electrode materials. Table 7 lists the electrosorption capacity, mass, and dimensions of electrode materials, and other basic information of all previous carbon electrode materials for the CDI systems. It can be observed that all researchers measured the electrosorption capacity by mg of NaCl per gram of the active material and they ignored measuring it by the geometric area or geometric volume. This leads to the comparison of all electrosorption capacities by the weight of the activated carbon materials in the electrode. However, our research focussed on improving the electrosorption capacity in terms of the geometric area and geometric volume. It can be seen that 3.63 wt% of a-SWCNT-coated RVC gave the highest amount of electrosorption capacity (10.40 mg/g) than that of the other electrode materials. Additionally, the electrode that had 23.58 wt% a-SWCNT-coated RVC still had a high electrosorption capacity (3.23 mg/g) compared to the CNT electrode material (2.33 mg/g). It is worth mentioning that the time required for one adsorption cycle in the CNT electrode material, which was 100 min, was decreased more than 16 times when the a-SWCNT-coated RVC electrode was used

(6 min). This means that the amount of pure water produced is increased more than 16 times when the a-SWCNT-coated RVC electrode is used.

Table 7. Comparison of electrosorption capacity of various carbon electrodes and electrodes developed in this article.

Parameters	Carbon Electrodes						Electrodes Developed in This Article (SWCNT/RVC) Using a New Cell			
	AC	CA	CNT	CNT/CNF	rGO	ACF	3.63 wt%	12.50 wt%	17.43 wt%	23.58 wt%
Electrosorption capacity (mg/g)	1.51	2.56	2.33	3.32	3.23	4.64	10.40	3.99	3.67	3.23
Electrosorption Capacity (mg/cm^3)	-	-	-	-	-	-	0.029	0.043	0.058	0.075
Electrosorption capacity (mg/cm^2)	-	-	-	-	-	-	0.039	0.057	0.078	0.10
Mass of materials (g)	1.50	4.30	0.12	0.85	1.50	0.31	0.006	0.023	0.034	0.050
Electrode length and width (cm)	14 × 7	16 × 8	8 × 8	8.7 Diameter	14 × 7	7 × 5	4 × 1.8			
Electrode thickness (mm)	0.3	0.8	0.03	0.3	0.3	0.2	3.0			
Applied voltage (V)	1.5	1.5	1.2	1.2	1.5	1.6	1.5			
Flow rate (mL/min)	20	400	40	14	20	5	50			
Solution volume (mL)	200	500	50	40	200	50	70			
Initial conductivity of NaCl solution (µS/cm)	145	101	100	100	145	192	143			
Time of electrosorption cycle (min)	30	840	100	45	30	150	6			
Reference	[24]	[25]	[22]	[23]	[24]	[21]				

The electrosorption capacity of the activated carbon nanofibre (ACF) web was consistently higher than that of AC, CA, CNT, CNT/CNF, rGO, and all a-SWCNT-coated RVC electrodes, except the 3.25 wt% SWCNT-coated RVC, as shown in Table 7. Unfortunately, this ACF electrode functioned with a very low flow rate (5 mL/min) and needed a long time to complete one cycle of adsorption (150 min). The main differences between the CDI electrodes include the energy consumed in the CDI system. It is known that the energy consumed increases with the increase in the time required for the adsorption processes and the increase in the cell voltage [38]. Table 7 shows the applied cell voltages and time for one adsorption cycle. It is noted that the applied cell voltage for all a-SWCNT-coated RVC electrodes was 1.5 V as for the AC, CA, and rGO electrodes. The differences between these electrodes were the time taken for one adsorption cycle (6 min for a-SWCNT-coated RVC electrode, 30 min for AC and rGO, and 840 min for CA). It is observed that the energy consumed in the CDI system with the a-SWCNT-coated RVC electrode is much less compared to the other electrodes. Additionally, a comparison of the time per adsorption cycle of a-SWCNT-coated RVC (6 min) with other electrodes in Table 7 shows that the time is much shorter, which leads to energy saving.

4. Conclusions

The construction of a unique 3D electrode based on a-SWCNT-coated RVC composites has been performed in this study and their electrochemical performance has been checked by cyclic voltammetry. The optimum voltage and flow rate for the electrodes were 1.5 V and 50 mL/min, respectively, at which the NaCl ion removal performance was mostly affected onto the sites of a-SWCNT. The effect of increasing a-SWCNT loading on the ion removal performance of a-SWCNT/RVC composite electrodes was investigated, which suggested that the composite electrodes can be effectively used in the CDI cell for desalination technology. The CDI system was tested for flow-between and flow-through electrode configurations. It is revealed that the flow-through electrode configuration takes 18 min for one desalination cycle, which means the desorption cycle is three times faster compared to the flow-between electrode configuration, which takes 54 min. A new CDI cell was designed based on the flow-through electrode configuration, which revealed that the electrosorption capacity for all electrodes was increased with the simultaneous reduction in desalination cycle time. The electrosorption capacity for the 3.63 wt% a-SWCNT-coated RVC electrode composite was increased by 23.96%, with a three times reduction in desalination cycle time when the flow-through electrode is used in a new CDI cell. It has been shown that when the distance between the two electrodes is increased, the adsorption time also increases. It is observed that the NaCl ion adsorption characteristics follow the pseudo-first-order kinetic model compared to the pseudo-second-order kinetic model. However, the electrosorption behavior of the 23.58 wt% a-SWCNT electrode is in good agreement with both the Langmuir and Freundlich isotherm models, suggesting monolayer adsorption as the primary mechanism in the electrosorption process. The maximum electrosorption capacity at the highest solution feed concentration was 8.89 mg/g, whereas the theoretical maximum value was 13.08 mg/g, as per the calculation using the Langmuir isotherm model. It has been shown that our developed electrode based on 3.63 wt% of a-SWCNT-coated RVC showed the highest amount of electrosorption capacity (10.40 mg/g) compared to the other electrode materials. Moreover, the time taken for one desalination cycle in the case of the CNT electrode materials (100 min) was more than 16 times less when our a-SWCNT-coated RVC electrode was used (6 min). This indicates that, within the same time period, the amount of pure water production increased more than 16 times when the a-SWCNT-coated RVC electrode is used. Additionally, the energy consumption in the CDI system with the a-SWCNT-coated RVC electrode is much less compared to the other electrodes.

Supplementary Materials: The following are available online at http://www.mdpi.com/2079-4991/8/7/527/s1. Figure S1: The full process of dip coating RVC in an a-SWCNT solution, Figure S2: Calibration curve linearity for ionic conductivity vs. NaCl concentration, Figure S3: Schematic diagram of a capacitive deionization (CDI) cell, Figure S4: (a–c) Photos and (d–f) SEM micrographs of 60, 45, and 30 ppi RVC samples, respectively; (g) the specific capacitance of a-SWCNT-coated RVC electrodes of various porosities in 1 M NaCl solution calculated from cyclic voltammograms recorded in a voltage range between -0.2 and 1.0 V using a three-electrode system vs. Ag/AgCl at a 5-mV/s scan rate, Figure S5: Scanning electron microscopy (SEM) images (a,b) top surface, (c) cross-section, and (d,e) 45° view of 23.58 wt% a-SWCNT loading, Figure S6: (a) Comparison cyclic voltammograms of the 1 cm^3 bare RVC electrode and the same size of the 23.58 wt% a-SWCNT-coated RVC electrode in 1 M NaCl scanned at 20 mV/s and using an RVC counter electrode and Ag/AgCl reference electrode in a three-electrode system; and (b) the effect of a-SWCNT loading of a-SWCNT/RVC composite electrodes on specific capacitance, Figure S7: (a) Cyclic voltammograms of the 3.63 wt% a-SWCNT-coated RVC electrode and (b) specific capacitance of various a-SWCNT-coated RVC electrodes at various scan rates in 1 M NaCl solution in the voltage range between -0.2 and 1.0 V vs. Ag/AgCl using a three-electrode system. Cyclic voltammograms of a-SWCNT-coated RVC electrodes at 20 mV/s, in terms of (c) current per gram of a-SWCNT and (d) current per geometric volume of the electrode; and (e) capacitance of the electrodes per gram of a-SWCNT and per geometric volume of the electrode, Table S1: Specific capacitance of various a-SWCNT-coated RVC electrodes by F/g of a-SWCNT, F/area of the electrode, and F/volume of the electrode, in 1 M NaCl solution at various scan rates.

Author Contributions: A.A. (Ali Aldalbahi), M.R., and M.A. have designed and performed the experiments and analyzed the data. A.A. (Abdulaziz Alrehaili) and A.H. have tested the samples. A.A. (Ali Aldalbahi), M.R., and M.A. have written the manuscript. All authors have read and approved the final manuscript.

Acknowledgments: The authors extend their appreciation to the Deanship of Scientific Research at King Saud University for funding this work through Research Group, No. RG-1436-005.

Conflicts of Interest: The authors declare no conflict of interest.

References

1. Ryoo, M.; Seo, G. Improvement in capacitive deionization function of activated carbon cloth by titania modification. *Water Res.* **2003**, *37*, 1527–1534. [CrossRef]
2. Yang, K.L.; Yiacoumi, S.; Tsouris, C. Electrosorption capacitance of nanostructured carbon aerogel obtained by cyclic voltammetry. *J. Electroanal. Chem.* **2003**, *540*, 159–167. [CrossRef]
3. Michael, M.S.; Prabaharan, S.R.S. High voltage electrochemical double layer capacitors using conductive carbons as additives. *J. Power Sources* **2004**, *136*, 250–256. [CrossRef]
4. Yang, C.M.; Choi, W.H.; Na, B.K.; Cho, B.W.; Cho, W.I. Capacitive deionization of NaCl solution with carbon aerogel-silicagel composite electrodes. *Desalination* **2005**, *174*, 125–133. [CrossRef]

5. Dai, K.; Shi, L.; Zhang, D.; Fang, J. NaCl adsorption in multi-walled carbon nanotube/active carbon combination electrode. *Chem. Eng. Sci.* **2006**, *61*, 428–433. [CrossRef]

6. Li, L.; Song, H.; Chen, X. Ordered mesoporous carbons from the carbonization of sulfuric-acid-treated silica/triblock copolymer/sucrose composites. *Microporous Mesoporous Mater.* **2006**, *94*, 9–14. [CrossRef]

7. Wang, S.; Wang, D.; Ji, L.; Gong, Q.; Zhu, Y.; Liang, J. Equilibrium and kinetic studies on the removal of NaCl from aqueous solutions by electrosorption on carbon nanotube electrodes. *Sep. Purif. Technol.* **2007**, *58*, 12–16. [CrossRef]

8. Zou, L.; Li, L.; Song, H.; Morris, G. Using mesoporous carbon electrodes for brackish water desalination. *Water Res.* **2008**, *42*, 2340–2348. [CrossRef] [PubMed]

9. Mayes, R.T.; Tsouris, C.; Kiggans, J.O., Jr.; Mahurin, S.M.; DePaoli, D.W.; Dai, S. Hierarchical ordered mesoporous carbon from phloroglucinol-glyoxal and its application in capacitive deionization of brackish water. *J. Mater. Chem.* **2010**, *20*, 8674–8678. [CrossRef]

10. Aldalbahi, A.; Rahaman, M.; Almoigli, M.; Meriey, A.; Alharbi, K.N. Improvement in electrode performance of novel SWCNT loaded three-dimensional porous RVC composite electrodes by electrochemical deposition method. *Nanomaterials* **2018**, *8*, 19. [CrossRef] [PubMed]

11. Chen, Z.; Song, C.; Sun, X.; Guo, H.; Zhu, G. Kinetic and isotherm studies on the electrosorption of NaCl from aqueous solutions by activated carbon electrodes. *Desalination* **2011**, *267*, 239–243. [CrossRef]

12. Li, H.; Pan, L.; Lu, T.; Zhan, Y.; Nie, C.; Sun, Z. A comparative study on electrosorptive behavior of carbon nanotubes and graphene for capacitive deionization. *J. Electroanal. Chem.* **2011**, *653*, 40–44. [CrossRef]

13. Le-Clech, P.; Chen, V.; Fane, T.A.G. Fouling in membrane bioreactors used in wastewater treatment. *J. Membr. Sci.* **2006**, *284*, 17–53. [CrossRef]

14. Jitianu, A.; Cacciaguerra, T.; Benoit, R.; Delpeux, S.; Béguin, F.; Bonnamy, S. Synthesis and characterization of carbon nanotubes-TiO$_2$ nanocomposites. *Carbon* **2004**, *42*, 1147–1151. [CrossRef]

15. Campidelli, S.; Klumpp, C.; Bianco, A.; Guldi, D.M.; Prato, M. Functionalization of CNT: Synthesis and applications in photovoltaics and biology. *J. Phys. Org. Chem.* **2006**, *19*, 531–539. [CrossRef]

16. Itoh, E.; Suzuki, I.; Miyairi, K. Field emission from carbon-nanotube-dispersed conducting polymer thin film and its application to photovoltaic devices. *Jpn. J. Appl. Phys.* **2005**, *44*, 636–640. [CrossRef]

17. Wang, L.; Wang, M.; Huang, Z.-H.; Cui, T.; Gui, X.; Kang, F.; Wang, K.; Wu, D. Capacitive deionization of NaCl solutions using carbon nanotube sponge electrodes. *J. Mater. Chem.* **2011**, *21*, 18295–18299. [CrossRef]

18. Yan, C.; Zou, L.; Short, R. Single-walled carbon nanotubes and polyaniline composites for capacitive deionization. *Desalination* **2012**, *290*, 125–129. [CrossRef]

19. Zhang, D.; Yan, T.; Shi, L.; Peng, Z.; Wen, X.; Zhang, J. Enhanced capacitive deionization performance of graphene/carbon nanotube composites. *J. Mater. Chem.* **2012**, *22*, 14696–14704. [CrossRef]

20. Li, L.; Li, F.; Xiao, Y.; Aigin, Z. The effect of carbonyl, carboxyl and hydroxyl groups on the capacitance of carbon nanotubes. *New Carbon Mater.* **2012**, *26*, 224–228. [CrossRef]

21. Wang, G.; Pan, C.; Wang, L.; Dong, Q.; Yu, C.; Zhao, Z.; Qiu, J. Activated carbon nanofiber webs made by electrospinning for capacitive deionization. *Electrochim. Acta* **2012**, *69*, 65–70. [CrossRef]

22. Nie, C.; Pan, L.; Li, H.; Chen, T.; Lu, T.; Sun, Z. Electrophoretic deposition of carbon nanotubes film electrodes for capacitive deionization. *J. Electroanal. Chem.* **2012**, *666*, 85–88. [CrossRef]

23. Wang, X.Z.; Li, M.G.; Chen, Y.W.; Cheng, R.M.; Huang, S.M.; Pan, L.K.; Sun, Z. Electrosorption of ions from aqueous solutions with carbon nanotubes and nanofibers composite film electrodes. *Appl. Phys. Lett.* **2006**, *89*, 053127. [CrossRef]

24. Wang, Z.; Dou, B.; Zheng, L.; Zhang, G.; Liu, Z.; Hao, Z. Effective desalination by capacitive deionization with functional graphene nanocomposite as novel electrode material. *Desalination* **2012**, *299*, 96–102. [CrossRef]

25. Jung, H.; Hwang, S.; Hyun, S.; Lee, K.; Kim, G. Capacitive deionization characteristics of nanostructured carbon aerogel electrodes synthesized via ambient drying. *Desalination* **2007**, *216*, 377–385. [CrossRef]

26. Aldalbahi, A.; Rahaman, M.; Almoiqli, M. A Strategy to enhance the electrode performance of novel three-dimensional PEDOT/RVC composites by electrochemical deposition method. *Polymers* **2017**, *9*, 157. [CrossRef]

27. Aldalbahi, A.; Rahaman, M.; Govindasami, P.; Almoiqli, M.; Altalhi, T.; Mezni, A. Construction of a novel three-dimensional PEDOT/RVC electrode structure for capacitive deionization: Testing and performance. *Materials* **2017**, *10*, 847. [CrossRef] [PubMed]

28. Li, H.; Zou, L.; Pan, L.; Sun, Z. Novel graphene-like electrodes for capacitive deionization. *Environ. Sci. Technol.* **2010**, *44*, 8692–8697. [CrossRef] [PubMed]

29. Kim, Y.; Hur, J.; Bae, W.; Choi, J. Desalination of brackish water containing oil compound by capacitive deionization process. *Desalination* **2010**, *253*, 119–123. [CrossRef]

30. Seo, S.; Jeon, H.; Lee, J.K.; Kim, G.-Y.; Park, D.; Nojima, H.; Lee, J.; Moon, S.-H. Investigation on removal of hardness ions by capacitive deionization (CDI) for water softening applications. *Water Res.* **2010**, *44*, 2267–2275. [CrossRef] [PubMed]

31. Suss, M.E.; Baumann, T.F.; Bourcier, W.L.; Spadaccini, C.M.; Rose, K.A.; Santiago, J.G.; Stadermann, M. Capacitive desalination with flow-through electrodes. *Energy Environ. Sci.* **2012**, *5*, 9511–9519. [CrossRef]

32. Pan, L.; Wang, X.; Gao, Y.; Zhang, Y.; Chen, Y.; Sun, Z. Electrosorption of anions with carbon nanotube and nanofibre composite film electrodes. *Desalination* **2009**, *244*, 139–143. [CrossRef]

33. Sfeir, M.; Beetz, T.; Wang, F.; Huang, L.; Huang, H.; Huang, M.; Hone, J.; Obrien, S.; Misewich, J.; Heinz, T.; et al. Optical spectroscopy of individual single-walled carbon nanotubes of defined chiral structure. *Science* **2006**, *312*, 554–556. [CrossRef] [PubMed]

34. Li, H.; Zou, L.; Pan, L.; Sun, Z. Using graphene nano-flakes as electrodes to remove ferric ions by capacitive deionization. *Sep. Purif. Technol.* **2010**, *75*, 8–14. [CrossRef]

35. Show, Y.; Imaizumi, K. Electric double layer capacitor with low series resistance fabricated by carbon nanotube addition. *Diam. Relat. Mater.* **2007**, *16*, 1154–1158. [CrossRef]

36. Mohrous, M.; Sakr, I.; Balabel, A.; Ibrahim, K. Experimental investigation of the operating parameters affecting hydrogen production process through alkaline water electrolysis. *Therm. Environ. Eng.* **2011**, *2*, 113–116. [CrossRef]

37. Nagai, N.; Takeuchi, M.; Kimura, T.; Oka, T. Existence of optimum space between electrodes on hydrogen production by water electrolysis. *Int. J. Hydrogen Energy* **2003**, *28*, 35–41. [CrossRef]

38. Andelman, M. Flow through capacitor basics. *Sep. Purif. Technol.* **2011**, *80*, 262–269. [CrossRef]

39. Hou, C.-H.; Huang, J.-F.; Lin, H.-R.; Wang, B.-Y. Preparation of activated carbon sheet electrode assisted electrosorption process. *J. Taiwan Inst. Chem. Eng.* **2012**, *43*, 473–479. [CrossRef]

40. Villar, I.; Suarez-De La Calle, D.J.; González, Z.; Granda, M.; Blanco, C.; Menéndez, R.; Santamaría, R. Carbon materials as electrodes for electrosorption of NaCl in aqueous solutions. *Adsorption* **2011**, *17*, 467–471. [CrossRef]

41. Malash, G.F.; El-Khaiary, M.I. Piecewise linear regression: A statistical method for the analysis of experimental adsorption data by the intraparticle-diffusion models. *Chem. Eng. J.* **2010**, *163*, 256–263. [CrossRef]

42. Li, H.; Lu, T.; Pan, L.; Zhang, Y.; Sun, Z. Electrosorption behavior of graphene in NaCl solutions. *J. Mater. Chem.* **2009**, *19*, 6773–6779. [CrossRef]

43. Li, H.; Pan, L.; Zhang, Y.; Zou, L.; Sun, C.; Zhan, Y.; Sun, Z. Kinetics and thermodynamics study for electrosorption of NaCl onto carbon nanotubes and carbon nanofibers electrodes. *Chem. Phys. Lett.* **2010**, *485*, 161–166. [CrossRef]

44. Li, H.; Zou, L. Ion-exchange membrane capacitive deionization: A new strategy for brackish water desalination. *Desalination* **2011**, *275*, 62–66. [CrossRef]

45. Boparai, H.K.; Joseph, M.; O'Carroll, D.M. Kinetics and thermodynamics of cadmium ion removal by adsorption onto nano zerovalent iron particles. *J. Hazard. Mater.* **2011**, *186*, 458–465. [CrossRef] [PubMed]

46. Senthil Kumar, P.; Gayathri, R. Adsorption of Pb2+ ions from aqueous solutions onto Bael tree leaf powder: Isotherms, kinetics and thermodynamics. *J. Eng. Sci. Technol.* **2009**, *4*, 381–399.

47. Purdom, P.W. *Environmental Health*, 2nd ed.; Academic Press: New York, NY, USA, 1980.

48. Marichev, V.A. Partial charge transfer during anion adsorption: Methodological aspects. *Surf. Sci. Rep.* **2005**, *56*, 277–324. [CrossRef]

nanomaterials

MDPI

Article

Titanium Dioxide Nanotubes as Model Systems for Electrosorption Studies

Xian Li [1], Samantha Pustulka [2], Scott Pedu [2], Thomas Close [2], Yuan Xue [3], Christiaan Richter [2,4] and Patricia Taboada-Serrano [1,2,*]

[1] Microsystems Engineering Ph.D. Program, Rochester Institute of Technology, Rochester, NY 14623-5603, USA; xl9206@rit.edu

[2] Department of Chemical Engineering, Rochester Institute of Technology, Rochester, NY 14623-5603, USA; spustulka3@gatech.edu (S.Pu.); spedu@andrew.cmu.edu (S.Pe.); tclose@mit.edu (T.C.); cpr@hi.is (C.R.)

[3] Materials Science Program, University of Rochester, Rochester, NY 14627-0216, USA; yxue4@ur.rochester.edu

[4] Industrial Engineering, Mechanical Engineering and Computer Science, School of Engineering and Natural Sciences, University of Iceland, Reykjavik 600169-2039, Iceland

* Correspondence: ptsche@rit.edu; Tel.: +1-585-475-7337

Received: 2 April 2018; Accepted: 23 May 2018; Published: 5 June 2018

Abstract: Highly ordered titanium dioxide nanotubes (TiO_2 NTs) were fabricated through anodization and tested for their applicability as model electrodes in electrosorption studies. The crystalline structure of the TiO_2 NTs was changed without modifying the nanostructure of the surface. Electrosorption capacity, charging rate, and electrochemical active surface area of TiO_2 NTs with two different crystalline structures, anatase and amorphous, were investigated via chronoamperometry, cyclic voltammetry, and electrochemical impedance spectroscopy. The highest electrosorption capacities and charging rates were obtained for the anatase TiO_2 NTs, largely because anatase TiO_2 has a reported higher electrical conductivity and a crystalline structure that can potentially accommodate small ions within. Both electrosorption capacity and charging rate for the ions studied in this work follow the order of $Cs^+ > Na^+ > Li^+$, regardless of the crystalline structure of the TiO_2 NTs. This order reflects the increasing size of the hydrated ion radii of these monovalent ions. Additionally, larger effective electrochemical active surface areas are required for larger ions and lower conductivities. These findings point towards the fact that smaller hydrated-ions experience less steric hindrance and a larger comparative electrostatic force, enabling them to be more effectively electrosorbed.

Keywords: electrosorption; titania nanotubes; nanostructured electrodes

---

1. Introduction

The properties of charged interfaces and the electrosorption of ions on charged surfaces have a remarkable influence on the kinetics of various electrochemical processes, including those having technological importance such as water desalination and energy storage [1–4]. Electrosorption is a process where ions of opposite charge (counter-ions) are immobilized within a region known as the electrical double layer (EDL), which forms in the vicinity of the electrolyte-electrode interface when a low direct current potential is applied [5,6]. Electrosorption takes place during electrochemical reactions, charging/discharging of batteries, operation of electrochemical capacitors (ECs) and capacitive deionization (CDI) [7,8]. Recently, CDI technology, which relies on the reversible removal of ions from the solution by trapping them within the EDL, has gathered great interest due to its low energy cost, environmental friendliness, and no secondary pollution [9,10]. Electrochemical capacitors (ECs) use the electric field in the EDLs established at the electrode-electrolyte interface to store electrical energy. Without Faradaic (redox) reaction at EC electrodes, the EDL energy storage mechanism makes fast energy uptake and delivery, and good power performance possible [11,12]. Therefore, the properties of the EDL crucially determine the ions that can be removed from aqueous solutions in CDI and the energy that can be stored in ECs. Finally, the characteristics of the EDL directly influence the outcome of electrochemical reactions by adding another "resistance" or providing more "housing" for electrostatic charge storage to the processes.

Previous literature on electrosorption mainly focused on electrode development, and was mostly for specific applications [13–15]. Porous carbons are often the choice of materials for electrosorption as they combine a high surface area and good electrical conductivity [16,17]. In spite of considerable efforts directed towards increasing surface area of carbon-based electrode materials in the last decades, CDI and ECs performance are still not at the required level in order to become competitive technologies. One of the factors contributing to this is the inaccessibility of the surface area and the disordered pore arrangements of carbon materials that result in only part of the total surface area being utilized during electrosorption. Additionally, it was found that electrosorption capacity is dependent on the type of ion and its characteristics, e.g., hydrated ion radius, charge and even specific interactions with the surface. These dependencies cannot be fully captured by the classical theory usually employed to model CDI and EC performance. Chen et al. noted that during CDI operation with activated carbon electrodes at the same concentration, smaller ions depicted size-affinity by being preferentially captured by EDLs [18]. Furthermore, they claimed that in a mixed Cl^- and NO_3^- solution, Cl^- is preferably electrosorped over NO_3^-. This interesting observation suggests that specific interactions between the ion and electrode indeed played a role in the electrosorption process since the Cl^- and NO_3^- have a similar hydrated radius (3.31 Å and 3.35 Å respectively), same concentration, and identical ionic charge in the test [18]. However, the current knowledge of this interaction is still rudimentary, and many experimental observations remain difficult to interpret because the property of carbon-based electrodes such as the architecture, composition, and pore size are difficult to control in order to perform a systematic study of these phenomena.

A model electrode with tunable properties is required in order to better understand the effects of electrode properties and their interactions with different ions on electrosorption. Electrodes with defined and tunable architectures, where pore sizes and pore lengths can be well defined, would allow for systematic investigation of the effects of confinement and surface properties on the electrosorption of ions with specific characteristics [19,20]. Owing to well-defined pore diameters and lengths via anodization of titanium foil in specifically formulated chemical baths; highly ordered and self-organized titanium dioxide nanotubes (TiO_2 NT) are a very appealing candidate for fulfilling this role [19,20]. A particular advantage of TiO_2 NT is that the crystalline structure is easily tuned by simply annealing between 350–450 °C in air (from amorphous to anatase), without modifying the electrode architecture. Anatase TiO_2 NT is a good approach to compensate for the low conductivity of amorphous TiO_2 NT, while offering a potentially different ion-accommodation mechanism. Anatase TiO_2 contains a cavity in its crystalline structure where a small ion could insert itself [21]. In the present work, TiO_2 NTs were fabricated and tested for their suitability as model electrodes for electrosorption studies. Electrosorption of three monovalent alkaline cations (Li^+, Na^+, Cs^+) was systematically explored in order to capture the interactive effects of the crystalline structure and hydrated ion radius on electrosorption capacity, charge dynamics and utilization of electrode surface area. The three ions chosen for the present work bear the same charge, but have distinctly different hydrated radiuses, which allowed us to investigate the effects of ion size during electrosorption. Additionally, the three ions selected are relevant in terms of their applications: desalination, energy storage and deionization of radioactive waste. The goal of the study was to demonstrate the suitability of anodized TiO_2 NT electrodes to study how electrode properties and ion characteristics affect electrosorption.

2. Materials and Methods

2.1. Fabrication of Titanium Dioxide Nanotube Electrodes

Titanium dioxide nanotube electrodes (TiO$_2$ NT) were fabricated following the method described in previous work [21,22]. Pieces of titanium foil (purity 99.7%, from STREM Chemicals Inc., Newburyport, MA, USA) were cut to the desired electrode size and rinsed with ethanol. A 150 mL solution of Ethylene Glycol and aqueous 0.05 mol/L ammonium fluoride with a volume ratio of 50:1 was introduced into a temperature-controlled anodization cell under dry air in order to synthesize electrodes at different anodization conditions. The foil was anodized at varying temperatures for 1 h in this solution with a platinum counter electrode at fixed voltages of 60 V. Temperature was one of the anodization parameters explored in order to define the electrode-surface architecture. Highly-ordered and uniform cylindrical tubes were the desired electrode architecture for the electrosorption studies. After anodization, the electrodes were removed from the solution, rinsed with methanol, and allowed to air-dry. Four electrode samples were prepared simultaneously in order to ensure consistent characteristics. Some of the samples were annealed in air at 425 °C for 1 h in order to change the TiO$_2$ crystalline structure from amorphous to anatase—this structure has a higher reported electrical conductivity [19]. The mass of the NTs was determined prior and after annealing by weighing, in order to detect any change in mass that might have occurred during annealing. Scanning electron microscopy (SEM, Zeiss Auriga CrossBeam SEM) was used to characterize the morphology of TiO$_2$-NT electrodes. The specific surface area was determined via statistical analysis of the dimensions of the nanotubes measured from SEM images taken at random locations on the electrodes synthesized for this work, in the same fashion as previous work [22]. The mass of the nanotubes was determined via direct weighing.

2.2. Electrochemical Tests

All electrochemical tests in this work were performed using a BASi Cell stand C3 potentiostat (Bioanalytical Systems BASi, West Lafayette, IN, USA), the counter electrode was a BASi MW-1032 platinum electrode, and the reference electrode was BASi Ag/AgCl (BASi MW-2052). The working electrodes were the TiO$_2$ NT prepared as detailed above. Electrochemical tests were performed with three monovalent alkaline metal ions (Li$^+$, Na$^+$, Cs$^+$) as counter-ions and the same co-ion (Cl$^-$) in 20 mL of each of the following 0.1 M solutions: Sodium Chloride (NaCl), Lithium Chloride (LiCl), and Cesium Chloride (CsCl). The area of the working electrode was 2 cm^2, corresponding to a mass of electrode (TiO$_2$ nanotubes and titanium foil) equal to 0.010 ± 0.001 g in all cases. The Ag/AgCl reference electrode and the platinum-wire counter electrode were placed in the solution along with the working electrode in a typical three-electrode setup. The volume and concentration of the solutions were selected to ensure that ions were not depleted from the bulk solution at higher applied potentials. This was to avoid ion-concentration effects on electrosorption capacity. A blanking procedure was run before each electrochemical test, comprising a constant potential of +600 mV vs. Ag/AgCl being held for 5 min. This was done to ensure that each test had the same initial conditions. The potentiostat was programmed for each electrochemical test with the specifications described below.

One chronoamperometry cycle (CA) included 2 s of quiet time with no applied potential (0 mV), 130 s at a constant negative applied potential (charge) and 130 s at the same constant positive applied potential (discharge). The pairs of negative and positive applied potentials during charge and discharge were −200 mV/+200 mV; −400 mV/+400 mV; and −600 mV/+600 mV. Each CA cycle (quiet time, charge, and discharge) was repeated at least fifteen times with a blanking procedure between each cycle in order to target true equilibrium electrosorption capacity with the measurements.

In the electrochemical impedance spectroscopy (EIS) test, the three-electrode cell was set up in the same way as for the CA and CV tests. The EIS tests were conducted with a Gamry Reference 600 (Gamry Instruments, Warminster, PA, USA) with the amplitude of the sinusoidal AC voltage signal of

5 mV over the frequency range 1 mHz to 1 kHz. Electrochemical impedance spectroscopy (EIS) spectra were analyzed with the Z-view software.

2.3. Electrosorption Capacity

Electrosorption capacity was calculated assuming that each electron unit of charge was neutralized by one ionic charge. The accumulated charge was calculated via numerical integration of the area underneath the current response curve with time during the charging cycle. Values for electrosorption capacity were double-checked via calculation of the charge released during the discharge part of the cycle.

3. Results

3.1. Surface Architecture of Anodized TiO2 Electrodes

Figure 1a,b shows SEM images (top and side view) of one of the annealed TiO_2 NT electrodes anodized with a potential of 60 V for 1 h at 15 °C. Regular nanotube structure with uniform diameters and lengths were obtained at these anodization conditions, which were subsequently used in electrosorption experiments. Furthermore, the as-prepared TiO_2 NT (non-annealed, amorphous) electrode is morphologically identical with the TiO_2 NT after annealing at 425 °C for 1 h in air. Statistical analysis of SEM images was used in order to determine the mean diameter and length for the TiO_2-NT electrodes prepared. Based on the measurement from the SEM images, it can be seen that the TiO_2 NT electrode has a diameter of 41.4 ± 4 nm, length of 2136 ± 50 nm and specific surface area of 31.4 ± 2.2 m^2/g. This level of control is highly desirable for electrosorption studies. However, it is also necessary to point out that surface architecture is highly sensitive to the anodization conditions. For instance, in Figure 1c, when anodization was carried out with the same anodization-bath formulation, applied potential and anodization time but at 5 °C, nano needle structures, instead of nanotubes, were obtained. This suggests that the low temperature may have changed the balance between the competitive oxidation reaction and oxide dissolution reaction taking place during Ti anodization in the presence of F^-. Another example is provided in Figure 1d, in which the anodization conditions were the same, but the temperature was 35 °C. One can observe that the nanotube structure is less regular and defined than the one depicted in Figure 1a. This is because the high temperature accelerates both the fluoride etching and oxide layer growth, resulting in the cluster-like nanotube bulk appearance and deconstruction of some nanotubes.

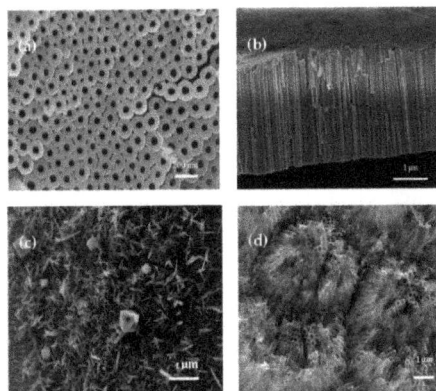

Figure 1. SEM images of TiO_2 NT electrodes (**a**), top view of TiO_2 NT anodized at 60 V 15 °C, annealed; (**b**) side view of TiO_2 NT anodized at 60 V 15 °C, annealed; (**c**) top view of TiO_2 NT anodized at 60 V 5 °C, non-annealed; (**d**) top view of TiO_2 NT anodized at 60 V 35 °C, non-annealed).

Electrodes of uniform TiO$_2$ NTs were prepared at applied potentials equal to 60 V, a temperature of 15 °C, and an anodizaion time of 1 h for subsequent electrosorption studies, in order to demonstrate their applicability as model electrodes. Half of the electrodes were annealed in order to modify their crystalline structure without modifying their architecture. The annealed electrodes are identified as TiO$_2$-NT-A, and the non-annealed electrodes are identified as TiO$_2$-NT-NA hereafter.

3.2. Effects of Electrode Crystalline Structure and Ionic Strength on Electrosorption

The effect of the TiO$_2$ NTs characteristic on electrosorption capacity was investigated in terms of its crystalline structure. As stated earlier, two groups of TiO$_2$ NTs: amorphous (non-annealed, TiO$_2$ NT-NA) and anatase (annealed, TiO$_2$ NT-A) were selected in order to study the effect of the TiO$_2$ crystalline structure on electrosorption capacity of similarly-charged ions. Anatase TiO$_2$ NT electrodes are prepared via annealing the as-synthesized amorphous electrodes in air at 350 °C for 1 h. The surface structure (TiO$_2$ NT diameter and length) of amorphous and anatase electrodes is identical, as it is not modified during annealing. Tighineanu et al. reported that the conductivity of anatase TiO$_2$ can be higher than that of amorphous TiO$_2$ by several orders of magnitude [19]. Due to high conductivity, it was expected that the TiO$_2$ NT-A would lead to high charge densities at the surface of the electrodes, i.e., larger electrosorption capacities. Figure 2a,b depicts electrosorption capacity for Cs$^+$, Na$^+$ and Li$^+$ for both TiO$_2$ NT-NA and TiO$_2$ NT-A electrodes. In Figure 2a, TiO$_2$ NT-A electrode shows higher electrosorption capacity for each ion than that of the TiO$_2$ NT-NA electrode, as expected. It is also shown that for TiO$_2$ NT-A, the highest electrosorption capacity, 29.78 μmol/m^2, was achieved during Cs$^+$ electrosorption at −600 mV. As for TiO$_2$ NT-NA, the highest electrosorption capacity was obtained for Cs$^+$ at −600 mV with a value equal to 12.19 μmol/m^2. This is only 40.9% of the highest electrosorption capacity of Cs$^+$ on TiO$_2$ NT-A at the same applied potential. Regardless of ion, the TiO$_2$ NT-A electrode substantially exhibited better electrosorption performance, which can be attributed to the higher conductivity of TiO$_2$-NT-A. Additionally, TiO$_2$-NT-A possesses a tunnel structure of anatase that may potentially aid electron transport and ion accommodation. It should be pointed out that, at a specific applied potential, the electrosorption capacity for various ions with both TiO$_2$ NT-A and TiO$_2$ NT-NA electrodes follows the order of Li$^+$ < Na$^+$ < Cs$^+$. For example, as seen in Figure 2b, with an applied potential of −600 mV, electrosorption capacities of Li$^+$, Na$^+$, and Cs$^+$ with TiO$_2$ NT-A were 17.88, 24.79, and 29.77 μmol/m^2, respectively, while with TiO$_2$ NT-NA, the electrosorption capacities were 6.82, 8.91, and 12.19 μmol/m^2 for Li$^+$, Na$^+$, and Cs$^+$, respectively. This trend can be explained by the fact that smaller hydrated ions experience less steric hindrance towards packing within the EDL and may also experience stronger electrostatic forces. In fact, the trend in electrosorption capacity aligns with the decreasing size of the hydration radii for Li$^+$, Na$^+$, and Cs$^+$ (which are equal to 3.82 Å, 3.58 Å, and 3.29 Å, respectively) [20]. Furthermore, the comparative advantage of smaller hydrated ions towards electrosorption is more marked with TiO$_2$ NT-A electrodes. For instance, at −200 mV, the elecrosorption capacity of Cs$^+$ with TiO$_2$ NT-NA was 2.9 times higher than that the corresponding value to Li$^+$, whereas this value was equal to 6.1 in the case of TiO$_2$ NT-A. The differences in electrosorption behavior did not limit themselves to capacity at equilibrium, but also in the dynamic behavior of the charging process (i.e., accumulation of charge within the EDL).

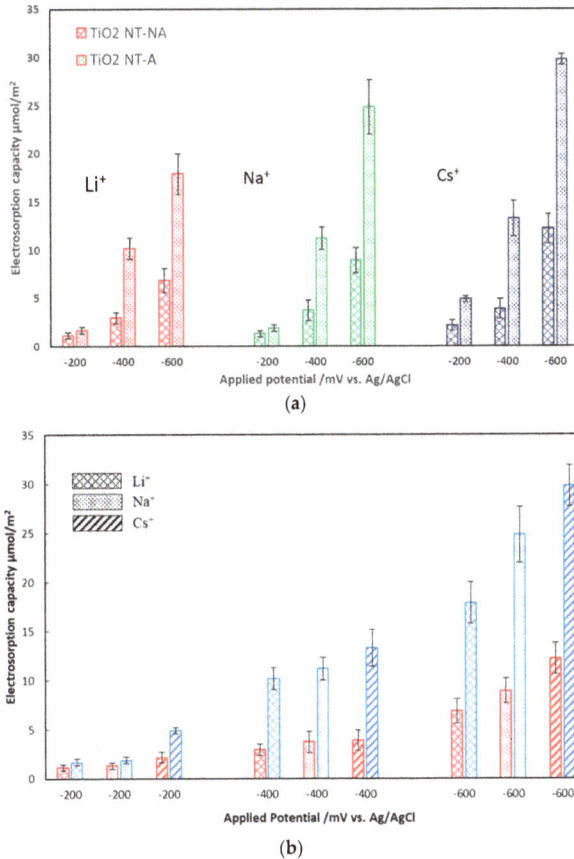

Figure 2. Electrosorption capacity of Li+, Na+, and Cs+ (a) at various applied potentials; and (b) for different electrodes (red bars for TiO$_2$ NT-NA and blue bars for TiO$_2$ NT-A).

3.3. Charging Dynamics Dependence on Ion and Crystalline Structure

Figure 3a–c shows the dynamic behavior of EDL formation for TiO$_2$ NT-A and TiO$_2$ NT-NA under −200 mV, −400 mV, and −600 mV, respectively. When the applied potential was −200 mV, the highest charging rate was obtained with the TiO$_2$ NT-A electrode and Cs+ in Figure 3a. Furthermore, it is evident that the charging rates for TiO$_2$ NT-A were consistently higher than the rates obtained for TiO$_2$ NT-NA with all the ions studied in this work. Increasing applied potential does not only increase the charging rate, but also accentuates the difference between charging rates of anatase and amorphous TiO$_2$ NT electrodes. When the applied potential increased to −400 mV, it became more apparent of the advantage of TiO$_2$ NT-A in terms of charging rate. From Figure 3b, two groups of charging curves are clearly identifiable resulting from the differences in the electrode crystalline structure. Figure 3c depicts even more dispersed behavior of the curves of charging rate, with a clear advantage for TiO$_2$ NT-A. One should notice that the charging rate is consistent with the decreasing hydrated ion radii. As seen in Figure 3a–c, the charging rate followed the order of Cs+ > Na+ > Li+ with both TiO$_2$ NT-A and TiO$_2$ NT-NA. Interestingly, the dispersion among charging curves brought about by the differences in hydrated ion radii become more marked as applied potential increased.

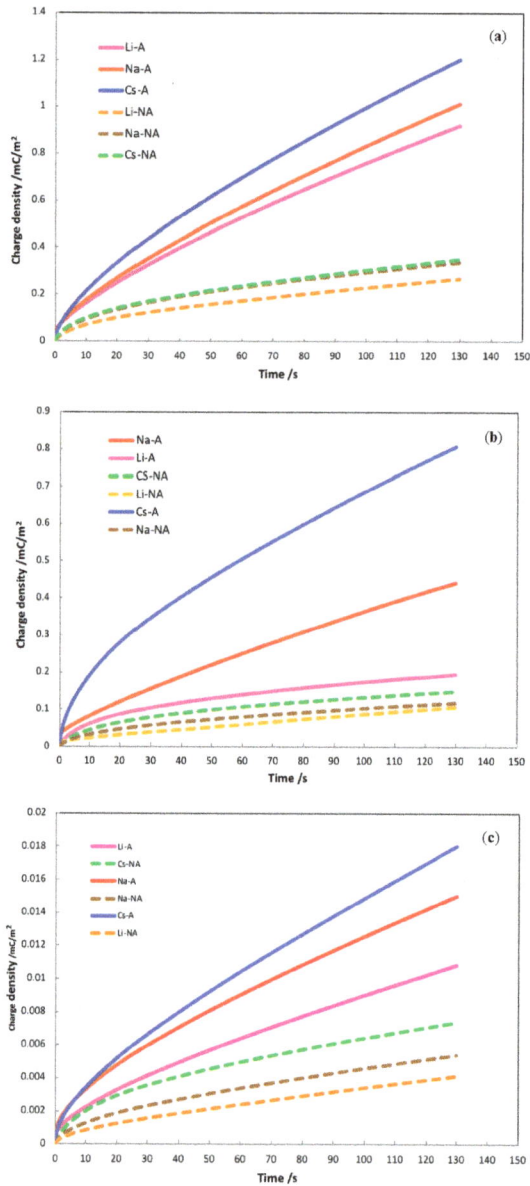

Figure 3. Charging rate of Li$^+$, Na$^+$, and Cs$^+$ with TiO$_2$-NT-A and TiO$_2$-NT-NA under (**a**) -600 mV vs. Ag/AgCl; (**b**) -400 mV vs. Ag/AgCl; and (**c**) -200 mV vs. Ag/AgCl (solid line for TiO$_2$-NT-A and dashed line for TiO$_2$-NT-NA).

Summarizing, (i) TiO$_2$ NT-A electrodes exhibited faster charging rates for each ion at every specific potential than that of the TiO$_2$ NT-NA electrode due to the high conductivity of anatase; (ii) Cs$^+$ with TiO$_2$ NT-A consistently exhibited the highest charging rates at each tested potential, followed by Na$^+$ and Li$^+$. This last finding confirms the fact that smaller hydrated ions have the advantage of less steric hindrance and most likely stronger electrostatic interactions during electrosorption.

3.4. Electrochemically Active Surface Area

It is not difficult to conclude that ion electrosorption benefits from the large surface area of TiO$_2$ NTs. The TiO$_2$ NT-A and TiO$_2$ NT-NA electrodes did not have any morphological difference, i.e., the surface structures on both electrodes were identical. Given the higher electrosorption capacity and charging rates exhibited by TiO$_2$ NT-A, it was necessary to determine how much of the total surface area of TiO$_2$ NTs was effectively used during electrosorption. Furthermore, it was necessary to explore how the TiO$_2$ NTs crystalline structure and any possible electrode-ion interactions may have influenced the effective area involved in electrosorption. Electrochemical active surface area (EASA) for each case was estimated based on cyclic voltammetry (CV) and electrochemical impedance spectroscopy (EIS) measurements [23–25]. In order to obtain EDL capacitance via CV, a series of CV tests at multiple scan rates were carried out. The current in the non-Faradaic potential region, measured via CV, which is assumed to be generated due to double-layer charging exclusively, is equal to the product of the scan rate, v, and the double-layer capacitance, C$_{DL}$, as given in Equation (1) [25,26].

$$i_{DL} = vC_{DL} \tag{1}$$

Examples of CVs of the TiO$_2$ NT-NA and TiO$_2$-A are depicted in Figure 4a,b. The CV tests were carried out in a non-Faradaic potential region at the following scan rates: 6, 12, 25, 50, and 100 mV/s. After acquiring the charging current at −0.3 V vs. Ag/AgCl, the cathodic and anodic charging currents were plotted as a function of scan rate v, which yields a straight line with a slope equal to C$_{DL}$, as depicted in Figure 4c,d. The determined electrochemical double-layer capacitance of the system, C$_{DL}$, is the average of the absolute value of the slope of the straight lines regressed from the experimental data.

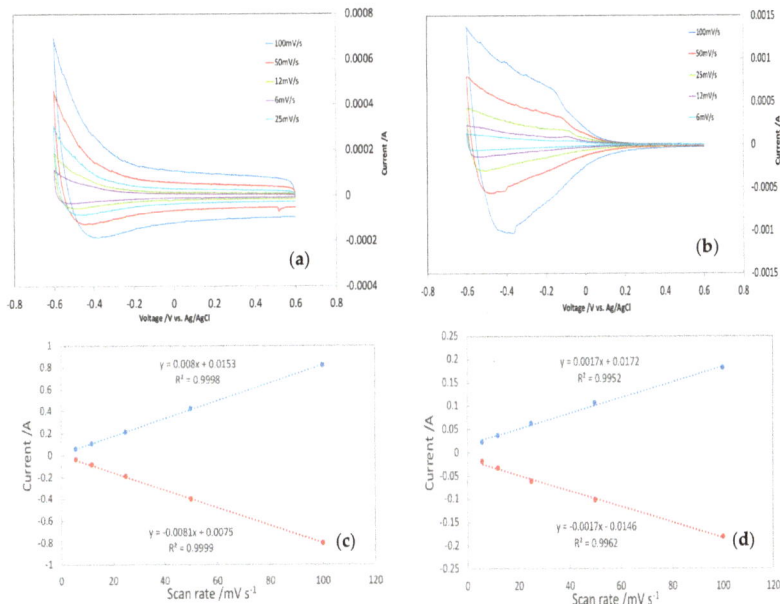

Figure 4. Cyclic voltammograms 1 M NaCl at various scan rates for (**a**) TiO$_2$ NT-NA; (**b**) TiO$_2$ NT-A; and EDL capacitance calculations for the determination of EASA (**c**) TiO$_2$ NT-NA; (**d**) TiO$_2$ NT-A. For the determination of EDL capacitance, charging currents measured at −0.3 V vs. Ag/AgCl, plotted as a function of the scan rate from Figure 4a,d, cathodic and anodic charging currents measured at −0.3 V vs. Ag/AgCl, plotted as a function of scan rate.

Upon having the double-layer capacitance, C_{DL}, the EASA of the sample was calculated according to Equation (2):

$$EASA = \frac{C_{DL}}{C_s \cdot m} \tag{2}$$

where m is the mass of TiO_2 NT, C_s is the specific capacitance of the sample and was calculated according to Equation (3):

$$C_s = \frac{C}{A} = \frac{Q}{E \times A}, \text{ where } Q = \int_0^t idt \tag{3}$$

where i is the time-dependent current response in the CA test, t is the time of charge/discharge in the CA test, E is the applied potential in the CA test, Q is the total transferred charge in the CA test, and A is the area of the electrode.

The double-layer capacitance was independently determined via electrochemical impedance spectroscopy (EIS). Figure 5a,b shows the Nyquist plots of different ions for TiO_2 NT amorphous (NA) and anatase (A), respectively. The Nyquist plots are interpreted with the help of an equivalent circuit, shown in the inset of Figure 5, in which the electrochemical system is approximated by the modified Randles circuit. The intersection of the impedance spectra with the real axis at the high frequency region end is the bulk resistance (R_s) including all contact resistance and resistance attributed to the electrolyte. The high frequency arc corresponds to the charge transfer limiting process and is ascribed to the charge transfer resistance (R_{ct}) at the contact interface between the electrode and electrolyte solution in parallel with a constant phase element (CPE) related to the double-layer capacitance. The frequency-dependent impedance of the CPE is given by Equation (4) [27,28]:

$$Z_{CPE} = \frac{1}{Q_0 \cdot (i\omega)^\alpha} \tag{4}$$

where Q_0 is a constant with dimensions $F\,s^{-(1-\alpha)}$, ω is the frequency of the sinusoidal applied potential, $i = (-1)^{1/2}$, and α is a dimensionless parameter, related to the phase angle of the frequency response, which has a value between 0 and 1 [27,28]. Based on the circuit model used here, the double-layer capacitance was calculated according to Equation (5) [27,28]:

$$C_{DL} = Q^{\frac{1}{\alpha}} \cdot \left[\left(\frac{1}{R_s} + \frac{1}{R_{ct}} \right) \right]^{(1 - \frac{1}{\alpha})} \tag{5}$$

Note that when $\alpha = 1$, the CPE behaves as a pure capacitor and $C_{DL} = Q$, and when $\alpha = 0$, the CPE behaves as a pure resistor and C_{DL} is not detectable [25,26]. For example, from the EIS measurement of the Li^+ with the annealed TiO_2 NT electrode at an applied potential of -0.05 V vs. Ag/AgCl shown in Figure 5, $R_s = 24.4\ \Omega$, $R_{ct} = 10.3\ \Omega$, $Q = 0.\ 148$ mF $s^{-(1-\alpha)}$, and $\alpha = 0.605$. The calculated C_{DL} from Equation (5) is 0.012 F. The calculated R_{ct} for all experimental conditions are listed in Table 1.

Table 1. Electrochemically-active surface area (EASA) for TiO_2 NT-A and TiO_2 NT-NA determined via CV and EIS.

Electrode	Target Ion	R_{ct}/Ω	EASA-CV/m²/g	EASA-EIS/m²/g	EASA-CV/Specific Area/%	EASA-EIS/Specific Area/%
TiO_2 NT-A	Li^+	10.3	14.1	13.7	44.7	43.6
TiO_2 NT-NA	Li^+	13.9	21.1	20.6	67.1	65.7
TiO_2 NT-A	Na^+	7.6	12.3	11.7	39.3	37.7
TiO_2 NT-NA	Na^+	9.9	17.8	16.8	56.8	57.7
TiO_2 NT-A	Cs^+	3.1	8.9	9.2	27.6	29.7
TiO_2 NT-NA	Cs^+	6.2	16.2	15.9	51.5	50.4

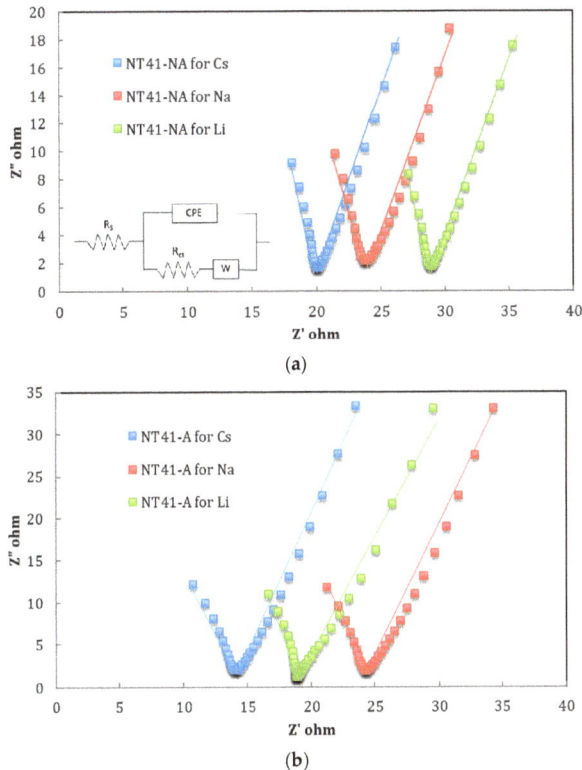

Figure 5. Nyquist plots for (**a**) TiO$_2$ NT-NA and (**b**) TiO$_2$ NT-A. The solid lines are the modeling fits to the EIS data by using the simplified Randles circuit shown in the inset of Figure 5a.

For a particular ion, the TiO$_2$ NT-A results in a lower charge transfer resistance (R$_{ct}$) than that of TiO$_2$ NT-NA, as expected, due to the higher electrical conductivity of the anatase phase. Two components are expected to contribute to the charge transfer resistance in electrosorption. The first one is due to the process of ion diffusion from the bulk electrolyte to the electrode surface. The second contributing factor is due to entrapment of ions within the EDL. The experimental conditions chosen for this work ensured that the ion concentration in the bulk electrolyte is much higher than the equilibrium concentration of electrosorption so that the resistance due to diffusion can be, in principle, neglected. Therefore, the charge transfer resistance can be equated to the resistance that an ion at the surface of an electrode has to overcome in order to be immobilized within the EDL. The R$_{ct}$ of different ions for both electrodes, TiO$_2$ NT-A and TiO$_2$ NT-NA, followed the trend of Li$^+$ > Na$^+$ > Cs$^+$, which is the reversed order of the hydrated ion radius as seen in Table 1. This finding confirmed our previous conclusion that the smaller hydrated ion radii presented lower steric resistance and possibly stronger electrostatic interactions. In general, the EDL capacitance measured by EIS is within 10% of that measured by the scan-rate dependent CVs, and the EASA obtained for a given sample by the two methodologies tend to agree within ±10%. Values of EASA were also correlated to hydrated-ion radius, but with a reverse trend: the largest ion Li$^+$ exhibits the largest EASA, followed by Na$^+$ and then Cs$^+$. It seems that packing of larger ions within the EDL requires more surface area. The EASA of 21.1 m^2/g, which is found in the Li$^+$ electrosorption test with TiO$_2$ NT-NA, is the largest value reaching 67.1% of the total surface area, while the smallest EASA of 8.9 m^2/g, 27.6% of the total surface area, is found in the Cs$^+$ electrosorption test with TiO$_2$ NT-A.

Summarizing, anodized TiO$_2$ NT electrodes of controlled surface structures proved to be a very effective model-electrode system in order to study experimentally ion-size effects and ion-electrode interaction effects during electrosorption.

4. Discussion

The crystalline structure of the TiO$_2$-NT electrode was a determinant factor of electrosorption capacity and charging rate for all ions studied in this work. Due to the high conductivity, the electrosorption capacity of the anatase TiO$_2$-NT electrode outperformed that of the amorphous TiO$_2$-NT electrode for all three targeted ions, Li$^+$, Na$^+$, and Cs$^+$ and tested potentials, −200 mV, −400 mV, and −600 mV. Additionally, the high conductivity of the anatase form of the TiO$_2$-NTs is favorable for fast EDL formation. As the surface architecture for all electrodes is identical, the only difference between the annealed and non-annealed electrodes was in the crystalline structure. This mainly affects the electrical conductivity. However, a tunnel structure in the annealed (anatase) electrodes is not present in the non-annealed electrodes. This tunnel structure could potentially play a role as well.

The electrosorption tests evidenced that the hydrated ion radius plays a critical role during electrosorption. The electrosorption and charging rate of both anatase and amorphous TiO$_2$-NT electrodes follow the same trend of Li$^+$ < Na$^+$ < Cs$^+$, which agrees with the order of decreasing hydrated ion radius Li$^+$ (3.82 Å) < Na$^+$ (3.58 Å) < Cs$^+$ (3.29 Å). The small ion can take advantage of low steric hindrance and higher relative electrostatic forces in order to be immobilized within the EDL more efficiently. This behavior becomes more marked with increasing applied potential. In fact, the lower charge transfer resistance is achieved by the combination of the anatase TiO$_2$ electrode and the ion with the smaller hydrated radius.

Higher electrical conductivity translates into higher effective potentials at the solid-liquid interface of the electrodes, i.e., larger driving forces during EDL charging and formation. They also translate into larger surface charge densities to be neutralized via EDL, i.e., higher electrosorption capacities. If EDL charging and structure only responded to ionic strength (ion concentration and ionic charge), as predicted by Classical EDL Theory, no differences should have been detected for ions bearing the same charge at the same concentrations. This work confirms predictions of molecular modelling that EDL structure is determined by competitive energy and steric effects [29–32]. This fact can also be visualized via the determination of EASA through independent electrochemical techniques.

The EASA based on cyclic voltammetry (CV) and EIS agree with each other very well. The amorphous TiO$_2$-NA electrode exhibited higher EASA than that of anatase TiO$_2$-NT, which is contrary to the electrosorption capacity for any particular ion at the same potential. It is noted that the amorphous TiO$_2$-NT utilized more surface area but reached a lower electrosorption capacity. The EASA of anatase TiO$_2$-NT decreased with the ion hydrated radius, which indicated that smaller hydrated ions are more easily packed within the EDL, confirming the observations gained by measuring electrosorption capacity and charging rate. Furthermore, the lower EASA in the case of TiO$_2$-NT-A confirms the fact that these electrodes present lower surface charge densities due to their lower electrical conductivity. The order of increasing EASA with increasing hydrated ion size (Cs$^+$ > Na$^+$ > Li$^+$) further confirms the occurrence of steric hindrance in the structure of the EDL.

5. Conclusions

A series of highly-ordered TiO$_2$ nanotubes (TiO$_2$-NT) with two different crystalline structures, amorphous and anatase, were synthesized by anodization, and were used as model-system electrodes in the study of ion size and crystalline structure effects during electrosorption of alkaline metal ions. The electrodes of controlled geometry allowed for the systematic investigation of ion size and crystalline effects on electrosorption capacity, charging dynamics, and electrochemical active surface area (EASA), via well-known electrochemistry techniques. The simple geometry of the electrodes allowed for the clear identification of steric effects on the behavior of EDL formation and structure that had been predicted via molecular simulations, but had not been unequivocally determined via experimentation. The highly tunable, regular nanotube structure of the electrodes proposed in this work, made the elucidation of steric effects during electrosorption and EDL formation possible.

Author Contributions: Conceptualization, P.T.-S. and X.L.; Methodology, P.T.-S., X.L. and C.R.; Validation, X.L., S.Pu., S.Pe., T.C. and Y.X.; Formal Analysis, X.L.; Investigation, X.L., S.Pu., S.Pe., T.C. and Y.X.; Resources, P.T.-S.; Data Curation, X.L., S.Pu., S.Pe., T.C. and P.T.-S.; Writing-Original Draft Preparation, X.L.; Writing-Review & Editing, X.L. and P.T.-S.; Visualization, X.L. and P.T.-S.; Supervision, P.T.-S.; Project Administration, P.T.-S.; Funding Acquisition, P.T.-S.

Funding: This research received no external funding.

Acknowledgments: To the Kate Gleason Fund.

Conflicts of Interest: The authors declare no conflict of interest.

References

1. Zhang, C.; He, D.; Ma, J.; Tang, W.; Waite, T.D. Faradaic reactions in capacitive deionization (CDI)-problems and possibilities: A review. *Water Res.* **2018**, *128*, 314–330. [CrossRef] [PubMed]

2. Khan, Z.U.; Yan, T.; Shi, L.; Zhang, D. Improved Capacitive Deionization by Using 3D Intercalated Graphene Sheet-Sphere Nanocomposite Architectures. *Environ. Sci. Nano* **2018**, *5*, 980–991. [CrossRef]

3. Yu, X.; Ma, H.; Du, W.; Qu, L.; Li, C.; Shi, G. Robust graphene composite films for multifunctional electrochemical capacitors with an ultrawide range of areal mass loading toward high-rate frequency response and ultrahigh specific capacitance. *Energy Environ. Sci.* **2018**, *11*, 559–565.

4. Wang, Z.; Tan, Y.; Yang, Y.; Zhao, X.; Liu, Y.; Niu, L.; Tichnell, B.; Kong, L.; Kang, L.; Liu, Z. Pomelo peels-derived porous activated carbon microsheets dual-doped with nitrogen and phosphorus for high performance electrochemical capacitors. *J. Power Sources* **2018**, *378*, 499–510. [CrossRef]

5. Foo, K.; Hameed, B. A short review of activated carbon assisted electrosorption process: An overview, current stage and future prospects. *J. Hazard. Mater.* **2009**, *170*, 552–559. [CrossRef] [PubMed]

6. Beden, B.; Morin, M.-C.; Hahn, F.; Lamy, C. "In situ" analysis by infrared reflectance spectroscopy of the adsorbed species resulting from the electrosorption of ethanol on platinum in acid medium. *J. Electroanal. Chem. Interfacial Electrochem.* **1987**, *229*, 353–366. [CrossRef]

7. Sun, Y.; Zeng, W.; Sun, H.; Luo, S.; Chen, D.; Chan, V.; Liao, K. Inorganic/polymer-graphene hybrid gel as versatile electrochemical platform for electrochemical capacitor and biosensor. *Carbon* **2018**, *132*, 589–597. [CrossRef]

8. Liu, P.; Yan, T.; Shi, L.; Park, H.S.; Chen, X.; Zhao, Z.; Zhang, D. Graphene-based materials for capacitive deionization. *J. Mater. Chem. A* **2017**, *5*, 13907–13943. [CrossRef]

9. Suss, M.; Porada, S.; Sun, X.; Biesheuvel, P.; Yoon, J.; Presser, V. Water desalination via capacitive deionization: What is it and what can we expect from it? *Energy Environ. Sci.* **2015**, *8*, 2296–2319. [CrossRef]

10. Liu, Y.; Nie, C.; Liu, X.; Xu, X.; Sun, Z.; Pan, L. Review on carbon-based composite materials for capacitive deionization. *RSC Adv.* **2015**, *5*, 15205–15225. [CrossRef]

11. Fic, K.; He, M.; Berg, E.J.; Novák, P.; Frackowiak, E. Comparative operando study of degradation mechanisms in carbon-based electrochemical capacitors with Li_2SO_4 and $LiNO_3$ electrolytes. *Carbon* **2017**, *120*, 281–293. [CrossRef]

12. Przygocki, P.; Abbas, Q.; Béguin, F. Capacitance enhancement of hybrid electrochemical capacitor with asymmetric carbon electrodes configuration in neutral aqueous electrolyte. *Electrochim. Acta* **2018**, *296*, 640–648. [CrossRef]

13. Rasines, G.; Lavela, P.; Macías, C.; Zafra, M.C.; Tirado, J.L.; Parra, J.B.; Ania, C.O. N-doped monolithic carbon aerogel electrodes with optimized features for the electrosorption of ions. *Carbon* **2015**, *83*, 262–274. [CrossRef]

14. Rasines, G.; Lavela, P.; Macías, C.; Zafra, M.; Tirado, J.; Ania, C. On the use of carbon black loaded nitrogen-doped carbon aerogel for the electrosorption of sodium chloride from saline water. *Electrochim. Acta* **2015**, *170*, 154–163. [CrossRef]

15. Li, K.; Guo, D.; Lin, F.; Wei, Y.; Liu, W.; Kong, Y. Electrosorption of copper ions by poly (m-phenylenediamine)/reduced graphene oxide synthesized via a one-step in situ redox strategy. *Electrochim. Acta* **2015**, *166*, 47–53. [CrossRef]

16. Kalluri, R.; Biener, M.; Suss, M.; Merrill, M.; Stadermann, M.; Santiago, J.; Baumann, T.; Biener, J.; Striolo, A. Unraveling the potential and pore-size dependent capacitance of slit-shaped graphitic carbon pores in aqueous electrolytes. *Phys. Chem. Chem. Phys.* **2013**, *15*, 2309–2320. [CrossRef] [PubMed]

17. Tsouris, C.; Mayes, R.; Kiggans, J.; Sharma, K.; Yiacoumi, S.; DePaoli, D.; Dai, S. Mesoporous carbon for capacitive deionization of saline water. *Environ. Sci. Technol.* **2011**, *45*, 10243–10249. [CrossRef] [PubMed]

18. Chen, Z.; Zhang, H.; Wu, C.; Wang, Y.; Li, W. A study of electrosorption selectivity of anions by activated carbon electrodes in capacitive deionization. *Desalination* **2015**, *369*, 46–50. [CrossRef]

19. Tighineanu, A.; Ruff, T.; Albu, S.; Hahn, R.; Schmuki, P. Conductivity of TiO$_2$ nanotubes: Influence of annealing time and temperature. *Chem. Phys. Lett.* **2010**, *494*, 260–263. [CrossRef]

20. Nightingale Jr, E. Phenomenological theory of ion solvation. Effective radii of hydrated ions. *J. Phys. Chem.* **1959**, *63*, 1381–1387. [CrossRef]

21. Richter, C.; Schmuttenmaer, C.A. Exciton-like trap states limit electron mobility in TiO$_2$ nanotubes. *Nat. Nanotechnol.* **2010**, *5*, 769–772. [CrossRef] [PubMed]

22. Li, X.; Close, T.; Pustulka, S.; Pedu, S.; Xue, Y.; Richter, C.; Taboada-Serrano, P. Electrosorption of monovalent alkaline metal ions onto highly ordered mesoporous titanium dioxide nanotube electrodes. *Electrochim. Acta* **2017**, *231*, 632–640. [CrossRef]

23. McCrory, C.C.; Jung, S.; Peters, J.C.; Jaramillo, T.F. Benchmarking heterogeneous electrocatalysts for the oxygen evolution reaction. *J. Am. Chem. Soc.* **2013**, *135*, 16977–16987. [CrossRef] [PubMed]

24. Benck, J.D.; Chen, Z.; Kuritzky, L.Y.; Forman, A.J.; Jaramillo, T.F. Amorphous molybdenum sulfide catalysts for electrochemical hydrogen production: Insights into the origin of their catalytic activity. *ACS Catal.* **2012**, *2*, 1916–1923. [CrossRef]

25. Bockris, J.M.; Srinivasan, S. Electrode kinetic aspects of electrochemical energy conversion. *J. Electroanal. Chem.* **1966**, *11*, 350–389. [CrossRef]

26. Boggio, R.; Carugati, A.; Trasatti, S. Electrochemical surface properties of Co$_3$O$_4$ electrodes. *J. Appl. Electrochem.* **1987**, *17*, 828–840. [CrossRef]

27. Orazem, M.E.; Tribollet, B. *Electrochemical Impedance Spectroscopy*; John Wiley & Sons, Inc.: Hoboken, NJ, USA, 2008; pp. 233–237.

28. Brug, G.; Van Den Eeden, A.; Sluyters-Rehbach, M.; Sluyters, J. The analysis of electrode impedances complicated by the presence of a constant phase element. *J. Electroanal. Chem. Interfacial Electrochem.* **1984**, *176*, 275–295. [CrossRef]

29. Taboada-Serrano, P.; Yiacoumi, S.; Tsouris, C. Behavior of mixtures of symmetric and asymmetric electrolytes near discretely charged planar surfaces: A Monte Carlo study. *J. Chem. Phys.* **2005**, *123*, 054703. [CrossRef] [PubMed]

30. Hou, C.-H. Monte Carlo simulation of electrical double-layer formation from mixtures of electrolytes inside nanopores. *J. Chem. Phys.* **2008**, *128*, 044705. [CrossRef] [PubMed]

31. Hou, C.-H.; Taboada-Serrano, P.; Yiacoumi, S.; Tsouris, C. Electrosorption selectivity of ions from mixtures of electrolytes inside nanopores. *J. Chem. Phys.* **2008**, *129*, 224703. [CrossRef] [PubMed]

32. Ney, E.M.; Hou, C.-H.; Taboada-Serrano, P. Calculation of electrical double layer potential profiles in nanopores from grand canonical Monte Carlo simulations. *J. Chem. Eng. Data* **2018**. [CrossRef]

nanomaterials

MDPI

Article

Electrodes Based on Carbon Aerogels Partially Graphitized by Doping with Transition Metals for Oxygen Reduction Reaction

Abdalla Abdelwahab [1,†], Jesica Castelo-Quibén [1], José F. Vivo-Vilches [1,‡], María Pérez-Cadenas [2], Francisco J. Maldonado-Hódar [1], Francisco Carrasco-Marín [1] and Agustín F. Pérez-Cadenas [1,*]

1   Carbon Materials Research Group, Department of Inorganic Chemistry, Faculty of Sciences, University of Granada, Campus Fuentenueva s/n, ES18071-Granada, Spain; abdalla.abdelwahab85@gmail.com (A.A.); jesicacastelo@ugr.es (J.C.-Q.); joseviv@ugr.es (J.F.V.-V.); fmaldon@ugr.es (F.J.M.-H.); fmarin@ugr.es (F.C.-M.)
2   Department of Inorganic and Technical Chemistry, Science Faculty, UNED, Paseo Senda del Rey 9, ES28040-Madrid, Spain; mariaperez@ccia.uned.es
*   Correspondence: afperez@ugr.es; Tel.: +34-958-243-316
†   Present address: Material Science and Nanotechnology Department, Faculty of Postgraduate Studies for Advanced Science, Beni-Suef University, Beni-Suef 62511, Egypt.
‡   Present address: Laboratoire de Chimie Physique et Microbiologie pour l'Environnement, UMR 7564 CNRS, 54600 Villers-lès-Nancy, France.

Received: 18 April 2018; Accepted: 20 April 2018; Published: 23 April 2018

Abstract: A series of carbon aerogels doped with iron, cobalt and nickel have been prepared. Metal nanoparticles very well dispersed into the carbon matrix catalyze the formation of graphitic clusters around them. Samples with different Ni content are obtained to test the influence of the metal loading. All aerogels have been characterized to analyze their textural properties, surface chemistry and crystal structures. These metal-doped aerogels have a very well-developed porosity, making their mesoporosity remarkable. Ni-doped aerogels are the ones with the largest surface area and the smallest graphitization. They also present larger mesopore volumes than Co- and Fe-doped aerogels. These materials are tested as electro-catalysts for the oxygen reduction reaction. Results show a clear and strong influence of the carbonaceous structure on the whole electro-catalytic behavior of the aerogels. Regarding the type of metal doping, aerogel doped with Co is the most active one, followed by Ni- and Fe-doped aerogels, respectively. As the Ni content is larger, the kinetic current densities increase. Comparatively, among the different doping metals, the results obtained with Ni are especially remarkable.

Keywords: carbon aerogel; graphitic cluster; metal nanoparticle; oxygen reduction reaction; electro-catalysis

1. Introduction

Nowadays the development of electric vehicles is one of the most promising alternatives to replace combustion engines and therefore there is a high interest in finding new environmental friendly sources of energy for automotive applications. For this reason, the production of electrical energy from chemical reactions by using fuel cells is a really interesting matter from both the industrial and fundamental research points of view [1–3]. Oxygen Reduction Reaction (ORR) takes place on the cathode in a fuel cell and several works can be found in the published literature about the synthesis and optimization of electro-catalytic materials for this reaction [1–11]. Of these, platinum-based electro-catalysts happen to be the most widely studied, since Pt is the most active metal for ORR [1,6–10]. Nevertheless, the rising price of platinum and other precious metals as Pd or Ir makes it more difficult to commercialize devices containing them. This is the reason why non precious metal electro-catalysts are more numerous and more studied in order to lower the costs of fuel cells [2,5–12].

On the other hand, carbon-based materials are being seriously considered as optimal candidates for ORR electro-catalysts [2,11,13,14]. Carbon gels are nanostructured materials, they are obtained from organic gels after their carbonization. Organic gels are prepared by polycondensation of organic monomers, normally resorcinol (R) and formaldehyde (F) [15]. The textural characteristics of carbon gels strongly depend on a precise control of the reactant concentrations and the conditions of the synthesis process: gelation, curing, drying and carbonization [16–18]. Surfactants can be added during the R-F polymerization which influences the morphology of the doped carbon gels and the metal dispersion [19]. Both surface area and pore volume, including the pore size distribution, are properties related to the synthesis conditions and processing that can be tuned, enabling the development of a wide set of materials with remarkable properties, e.g., for adsorption [20,21], catalysis [22–25] and electrochemical applications [26,27]. Besides, carbon gels doped with transition metals exhibit a homogeneous distribution together with a high dispersion of the metals throughout the carbon matrix [28,29]. Thus, transition metals are used in order to be anchored into the carbon matrix, which minimizes their leaching in liquid phase applications [19].

In the present work, carbon aerogels doped with iron, cobalt, and nickel were prepared, including a different nickel loading, exhaustibly characterized from textural, chemical, and electro-chemical points of view. Finally, their performances as electro-catalysts for the oxygen reduction reaction were evaluated and discussed in terms of their differences in porous texture, chemical characteristics, and metal doping, being the results of this comparative study the main objective of this work.

2. Materials and Methods

2.1. Preparation and Characterization of the Materials

Carbon aerogels doped with Ni, Co and Fe were prepared from resorcinol (R) and formaldehyde (F) dissolved in water (W) and using nickel, cobalt, or iron acetate as a catalyst precursor (C). The molar ratios were R:F = 1:2 and R:W = 1:17. Fe- and Co-doped aerogels were prepared only with a 6 wt %, approx. of metal loading while Ni-doped aerogels were prepared with 1, 4 and 6 wt %, approx. varying the amount of C. When an organic sol-gel solution was obtained it was cast into glass molds. After that, the curing process to obtain the organic gels was: 1 day at 40 °C, and 5 days at 80 °C. The glass molds were broken, and the organic aerogels were immersed in acetone for 24 h. Finally, the organic aerogels were treated with supercritical CO_2 for their drying. Another sample (A0), to be used as reference, was also prepared but without any transition metal. The organic aerogels were carbonized at 900 °C to obtain the carbon gels using a N_2 flow and a heating rate of 1 °C min^{-1}. The obtained carbon aerogels (A) were named as: ANi1, ANi4, ANi6, AFe6 and ACo6, the numbers indicate the approximate metal content in percentage. The metal loadings of the aerogels were determined by burning off a portion of a sample at 900 °C in air and weighting the residue.

The aerogels were texturally characterized by gas adsorption, scanning electron microscopy (SEM), and high-resolution transmission electron microscopy (HRTEM), and chemically characterized by X-ray diffraction (XRD), Raman spectroscopy and X-ray photoelectron spectroscopy (XPS). Samples performance for Oxygen Reduction Reaction was tested by means of cyclic voltammetry (CV) and linear sweep voltammetry (LSV).

N_2 and CO_2 adsorptions were carried out at −196 °C and 0 °C, respectively. Prior to measuring, the samples were outgassed for 24 h at 110 °C under high vacuum (10^{-6} mbar). The BET equation was applied to the N_2 adsorption obtaining the apparent surface area, S_{BET}. The Dubinin-Radushkevich (DR) equation was applied to the N_2 and CO_2 adsorption data and the corresponding micropore volume (W_0) and micropore mean width (L_0) were obtained. Total pore volumes ($V_{0.95}$) were determined from the N_2 adsorption isotherms at −196 °C and at 0.95 relative pressure. Finally, the mesopore volumes (V_{BJH}) and the mean mesopore widths (L_{BJH}) were obtained applying the BJH method [30] to the desorption branch of the N_2 isotherms.

SEM was carried out using a Zeiss SUPRA40VP scanning electron microscope (Carl Zeiss AG, Oberkochen, Germany), equipped with a secondary electron detector, back-scatter electron detector, and using a X-Max 50 mm energy dispersive X-ray microanalysis system. All the samples were crushed before realizing this analysis.

HRTEM was performed using a FEI Titan G2 60-300 microscope (FEI, Eindhoven, The Netherlands) with a high brightness electron gun (X-FEG) operated at 300 kV and equipped with a Cs image corrector (CEOS) and for analytical electron microscopy (AEM) a SUPER-X silicon-drift window-less EDX detector. The AEM spectra were collected in STEM (scanning transmission electron microscopy) mode using a HAADF (high angle annular dark field) detector. Digital X-ray maps were also collected on selected areas of the samples.

Raman spectra were recorded using a Micro-Raman JASCO NRS-5100 dispersive spectrometer (JASCO Inc, Easton, MD, USA) with a 532 nm laser line. From these spectra the ratio I_G/I_D was calculated as the quotient between the maximum intensity of each band.

XRD analysis was carried out with BRUKER D8 ADVANCE diffractometer (BRUKER, Rivas-Vaciamadrid, Spain) using CuK radiation. JCPDS files were used to assign the different diffraction peaks observed. Diffraction patterns were recorded between $10°$ and $70°$ (2θ) with a step of $0.02°$ and a time per step of 96 s. The average crystal size (d_{XRD}) was determined using the Scherrer equation.

XPS measurements of the carbon aerogels were performed using a Physical Electronics ESCA 5701 (PHI, Chanhassen, MN, USA) equipped with a MgK X-ray source ($h\nu = 1253.6$ eV) operating at 12 kV and 10 mA, and a hemispherical electron analyzer. The obtained binding energy (BE) values were referred to the C_{1s} peak at 284.7 eV. A base pressure of 10^{-9} mbar was maintained during data acquisition. Survey and multi-region spectra were recorded at C_{1s}, O_{1s}, Fe_{2p}, Co_{2p} and Ni_{2p} photoelectron peaks. Each spectral region was scanned enough times to obtain adequate signal-to-noise ratios. The spectra obtained after a background signal correction were fitted to Lorentzian and Gaussian curves to obtain the number of components, the position of each peak and the peak areas.

2.2. Electro-Chemical Studies. Oxygen Reduction Reaction

Cyclic Voltammetry (CV) and Linear Sweep Voltammetry (LSV) experiments were conducted on a three-electrode cell controlled by a Biologic VMP multichannel potentiostat (Bio-Logic Spain, Barcelona, Spain). A Rotating Disk Electrode (RDE) Metrohm AUTOLAB RDE-2 with a 3 mm Glassy Carbon tip (Gomensoro S.A, Madrid, Spain) was used as a working electrode. 5 mg of electro-catalyst were suspended on 1 mL of a solution which contained Nafion (5%) and water in a 1:9 (*v:v*) ratio. Subsequently, 10 μL of this suspension were loaded on RDE tip and dried under an infrared lamp [14]. The glassy carbon electrode had been previously polished with 1, 0.3 and 0.05 μm alumina powder and sonicated in deionized water and ethanol. Ag/AgCl was chosen as a reference electrode and Pt-wire as a counter electrode. The three electrodes were immersed in a 0.1 M KOH (electrolyte) solution in water.

The oxygen reduction reaction may occur by two different pathways: one implies 2 e_s^- transference and the formation of peroxide species (Equation (1)) which could damage the electro-catalytic layer which is not desirable; the other leads only to the formation of hydroxide and it occurs by a 4 e_s^- (Equation (2)) which is the requested one.

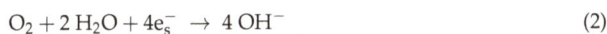

$$O_2 + H_2O + 2e_s^- \rightarrow HO_2^- + OH^- \tag{1}$$

$$O_2 + 2\,H_2O + 4e_s^- \rightarrow 4\,OH^- \tag{2}$$

CV experiments were carried out while N_2 or O_2 bubbled through the electrolyte solution during the measurements. The chosen potential window ranged from -0.8 to 0.4 V (at 5 mV·s^{-1} and 50 mV·s^{-1}). LSV curves were obtained in O_2-saturated 0.1 M KOH solutions at a different rotation speed and sweeping voltage, from 0.4 to -0.8 V (5 mV·s^{-1}). Data were fitted to the Koutecky-Levich model (Equations (3) and (4)) in order to evaluate the electro-catalytic performance of the samples and the transferred electron number for each of them [14].

$$\frac{1}{j} = \frac{1}{j_k} + \frac{1}{B\omega^{0.5}} \tag{3}$$

$$B = 0.2nF(D_{O_2})^{2/3}v^{-1/6}C_{O_2} \tag{4}$$

where j, current density; j_k, kinetic current density; ω, rotation speed; F, Faraday constant; D_{O_2}, oxygen diffusion coefficient (1.9×10^{-5} cm^2·s^{-1}); v, viscosity (0.01 cm^2·s^{-1}); C_{O_2}, oxygen concentration (1.2×10^{-6} mol·cm^{-3}).

3. Results

Table 1 collects the names and the textural properties of the samples. All carbon aerogels are microporous and mesoporous materials with significant mesopore volumes and BET surfaces areas. Aerogels doped with Ni have the highest surface areas and pore volumes among the metal-doped samples, specially the highest micropore volumes; among the Ni samples ANi6 is the most microporous material. In the opposite site, AFe6 has the lowest surface area and pore volumes. Aerogel A has textural properties comparable with the rest of samples.

Figure 1 shows the morphology of the samples studied by SEM. The structure of the carbon gels consist in a network formed by rounded particles with a different degree of fusion [31]; a very well-developed macroporous structure is also observed. No significant morphological differences are observed by SEM among the samples.

Table 1. Name, surface areas and pore volumes of the doped carbon gels.

Sample	S_{BET} m^2·g^{-1}	W_0 (N$_2$) cm^3·g^{-1}	L_0 (N$_2$) nm	W_0 (CO$_2$) cm^3·g^{-1}	L_0 (CO$_2$) nm	$V_{0.95}$ (N$_2$) cm^3·g^{-1}	V_{BJH} (N$_2$) cm^3·g^{-1}	L_{BJH} nm
A0	700	0.276	1.20	0.249	1.06	1.21	0.89	19.8
ANi1	663	0.258	1.07	0.276	0.63	0.82	0.54	17.1
ANi4	685	0.268	0.96	0.280	0.63	0.71	0.46	16.9
ANi6	698	0.273	0.90	0.294	0.64	0.69	0.48	16.8
ACo6	589	0.230	1.00	0.181	0.57	0.65	0.40	14.1
AFe6	461	0.177	1.00	0.182	0.62	0.41	0.25	12.3

Regarding the metal phase characterization, HRTEM analysis indicate that metals are mainly embedded within the carbon matrix; metal particles are clearly shown in Figures 2 and 3, and these are very well dispersed throughout the aerogel texture. Moreover, metal particles are detected within a wide range of nanometric sizes (Figure 4). On the other hand, these metal nanoparticles have catalyzed a partial graphitization around them during the pyrolysis; this fact was observed in all cases. Good examples of these graphitic clusters are shown in Figure 2, aerogels ANi6 and AFe6. It should be noted that the above-mentioned graphitization was not observed in sample A by HRTEM. These graphite clusters in the doped carbon aerogel structure were also observed in other works [31,32] with Co, Fe and Ni as doping metals. This can also be detected by XRD as a wide signal at around 26 θ (peak 002 of graphite, JCPDS card No. 41-1487) specially in the case of aerogels ACo6 and AFe6 (Figure 5), although this signal hardly can be observed in the Ni doped samples. This would indicate that the graphitic clusters in the Ni samples probably have mean crystallite sizes smaller than 4 nm or a very thin laminar form.

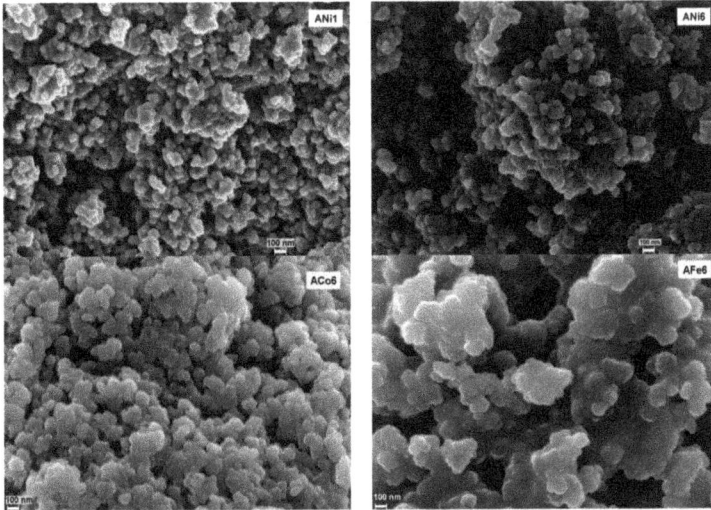

Figure 1. SEM microphotographs obtained at 100.00 KX of magnification of the samples ANi1, ANi6, ACo6 and AFe6.

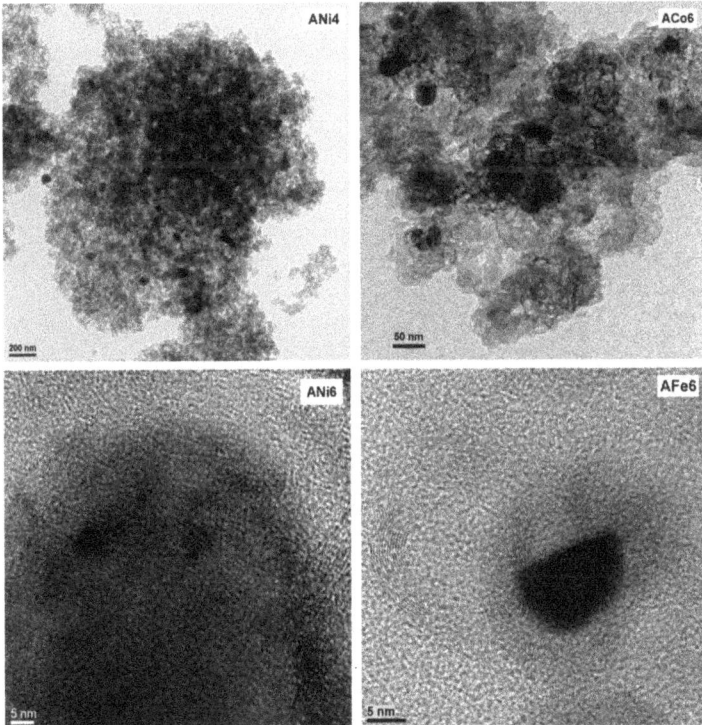

Figure 2. HRTEM images of the samples ANi4, ACo6, ANi6 and AFe6.

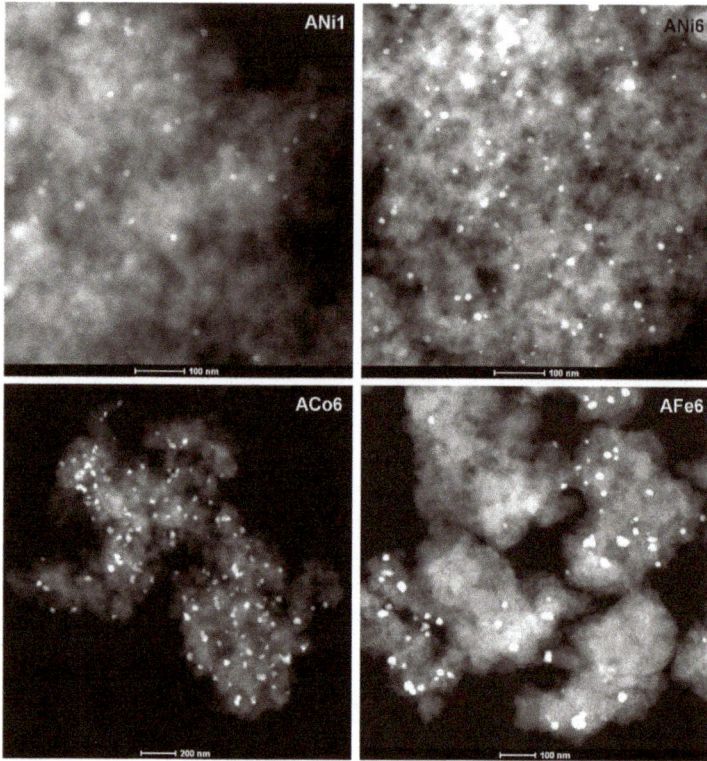

Figure 3. AEM spectra collected in STEM mode using a HAADF detector of the samples ANi1, ACo6, ANi6 and AFe6.

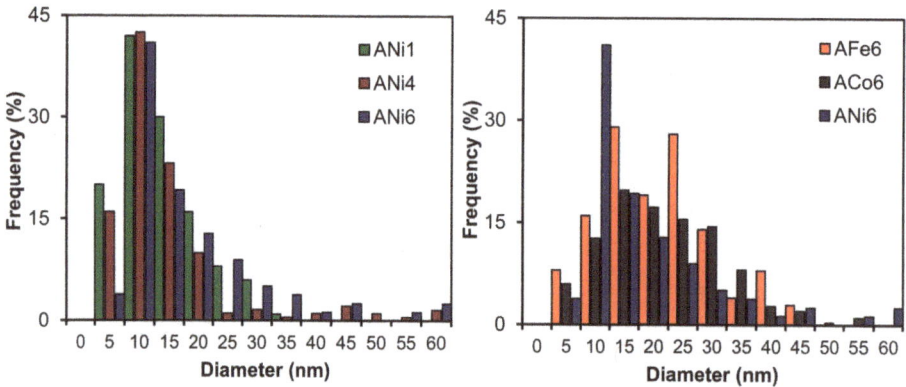

Figure 4. Particle size distributions obtained from HRTEM images.

On the other hand, the XRD peaks in Figure 5 clearly show the presence of Ni and Co completely reduced (JCPDS cards No. 04-0850, and 15-0806, respectively). Only in the case of sample AFe6 a mixture of Fe (0) (peaks at 44.6° and 65.1°, (JCPDS card No. 06-0696)) and Fe (III) (at 43.5°) could be detected; although the coincidence of this signal with the peak (101) of graphite makes both its assignation and resolution, difficult.

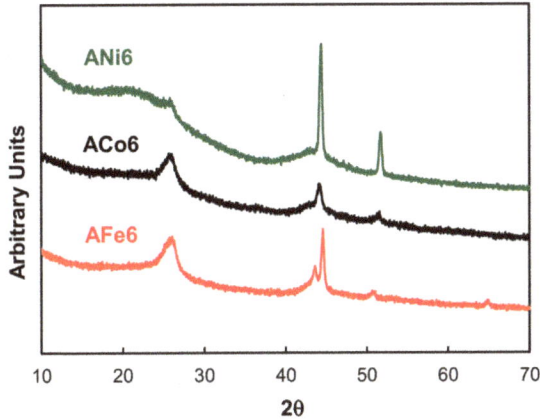

Figure 5. XRD patterns of aerogels ANi6, ACo6 and AFe6.

Analyzing the XP spectra, the peaks corresponding with metal phases cannot be practically distinguished from the base line in the case of ANi1 and ANi4, this means that Ni concentration on the external surface of these samples can be considered negligible. Only Ni2p, Co2p and Fe2p spectra of aerogels with 6 wt % could be analyzed. Figure 6 shows in the Ni2p spectrum only one Ni2p$_{3/2}$ signal at 853.3 eV which is assigned to Ni (II) [33]; its corresponding satellite peak can be clearly observed at 859.8 eV. In this line, only one Co2p$_{3/2}$ signal is observed at 781.1 eV of BE together its corresponding satellite at 786.1 eV, which is also assigned to Co (II) species [33]. Finally, the Fe2p spectrum contains two species of iron at 710.7 and 712.7 eV being these signals assigned to Fe$_2$O$_3$ (76.5%) and Fe$_3$O$_4$ (23.5%), respectively [33].

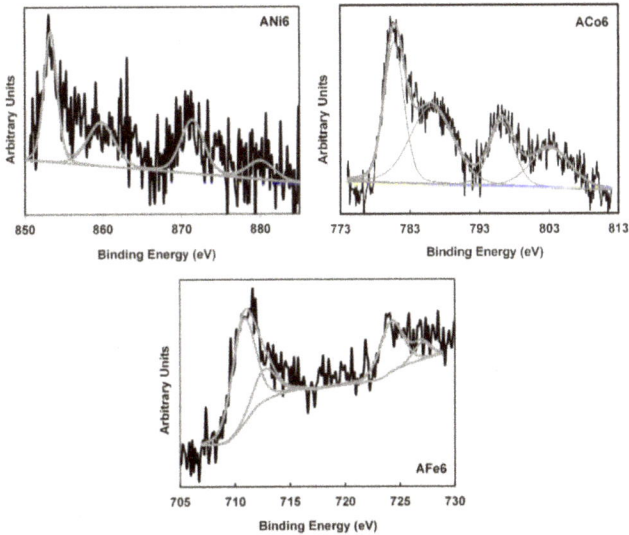

Figure 6. XP spectra of the doped carbon aerogels.

Raman spectra show (Figure 7) two main peaks at 1340 and 1580 cm^{-1} approx. which correspond to the D and G bands respectively [33]. In carbon aerogels, the D band can be associated with alternating ring vibrations in condensed benzene rings [34], while the G band can be associated with the development of the sp^2 carbon structure throughout the material during the carbonization process. It should be noted that carbon gels are normally amorphous carbon materials. Besides this, the intensity of the G band (I_G) with respect to its D band (I_D) is higher in the Ni doped aerogels than in the case of Fe or Co samples, and among the Ni samples this ratio I_G/I_D is clearly higher in ANi6 and ANi4 than in ANi1 (Table 2).

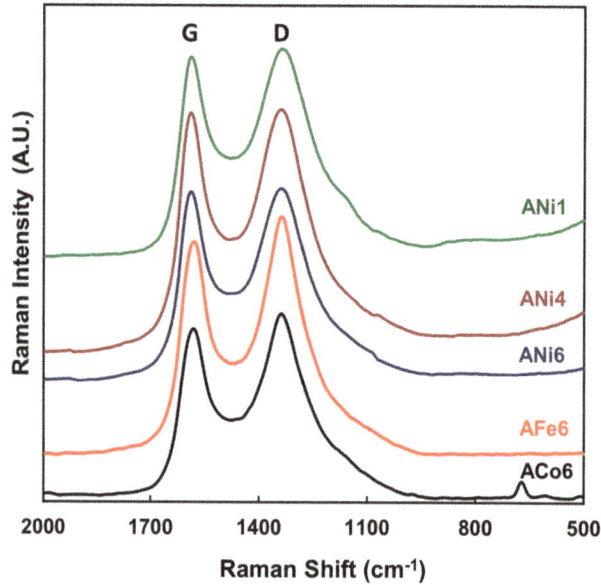

Figure 7. Raman spectra of the doped carbon aerogels.

Table 2. Chemical characteristics of the carbon aerogels.

Sample	Metal$_{TOTAL}$	Metal$_{XPS}$	O$_{XPS}$	d$_{XRD}$	d$_{HRTEM}$	I$_G$/I$_D$
	wt %	wt %	wt %	nm	nm	
A0	n.d.	n.d.	1.4	n.d.	n.d.	-
ANi1	1.2	n.d	1.6	15.9	11.9	0.97
ANi4	3.9	n.d	1.6	17.4	15.5	0.99
ANi6	5.8	0.3	1.8	21.1	17.7	0.99
ACo6	5.9	0.7	3.6	21.5	19.4	0.92
AFe6	6.1	0.4	2.9	21.6	18.6	0.89

n.d: no detected.

Table 2 collects the metal crystallite sizes estimated by applying the Scherrer equation, the mean particle sizes obtained from HRTEM, the I_G/I_D ration obtained from Raman spectra, the chemical composition obtained by XPS and total metal content of the aerogels. Among the Ni samples, the mean nickel particle size clearly increases with the metal loading; however, the samples ANi6, ACo6 and AFe6 show a very similar value around 21 nm.

Regarding the Rotating Disk Electrode (RDE) experiments, cyclic voltammetry was used in order to observe the difference between the samples behavior on a N$_2$-saturated electrolyte (KOH 0.1 M) and an O$_2$-saturated one. Figure 8 shows CV curves for ANi6 sample at 5 mV·s^{-1} and at 50 mV·s^{-1}, as well as for AFe6 and ACo6 samples at 50 mV·s^{-1} for comparison. In all cases a peak corresponding to the oxygen reduction can be observed when the curve is obtained on the O$_2$-saturated electrolyte.

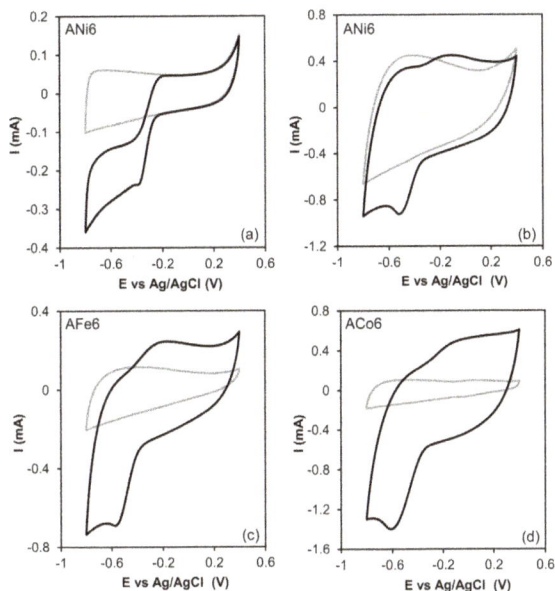

Figure 8. Cyclic voltammetries on N$_2$-saturated KOH 0.1 M (grey) and O$_2$-saturated KOH 0.1 M (black). (a) 5 mV·s^{-1}; (b–d) 50 mV·s^{-1}.

After CV, the electro-catalytic performance of the samples for oxygen reduction was studied by Linear Sweep Voltammetry (LSV). The experiments were conducted at a different rotating speed to apply the Koutecky-Levich Equation. This analysis is shown in Figure 9 for the ANi6 sample.

From this analysis the number of electrons transferred at a given potential can be obtained (Table 3). Aerogels with different content in Ni were tested to analyze the influence of the metal content in the electro-catalytic behavior of the samples on LSV (Figure 10a) and the number of electrons transferred (Figure 10b). Finally, aerogels doped with the three different metals but with the same metal loading (ANi6, ACo6 and AFe6) were compared as well (Figure 11). None catalytic activity was detected with the un-doped aerogel A0, neither by CV nor LSV.

Table 3. Parameters obtained from the analysis of LSV curves (values of *n* refer to K-L fitting for data at −0.8 V).

Sample	E_{onset}	j_k	n
	V	mA·cm^{-2}	
ANi1	−0.22	16.6	3.1
ANi4	−0.21	16.9	3.9
ANi6	−0.21	28.1	4.2
ACo6	−0.17	34.9	3.6
AFe6	−0.22	26.2	4.1

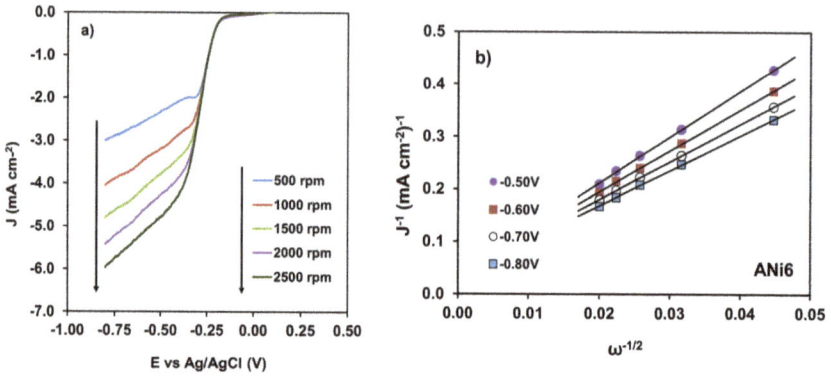

Figure 9. (**a**) LSV for ANi6 at different RDE rotating speed. (**b**) Koutecky-Levich fits at different potentials: from −0.5 to −0.8 V.

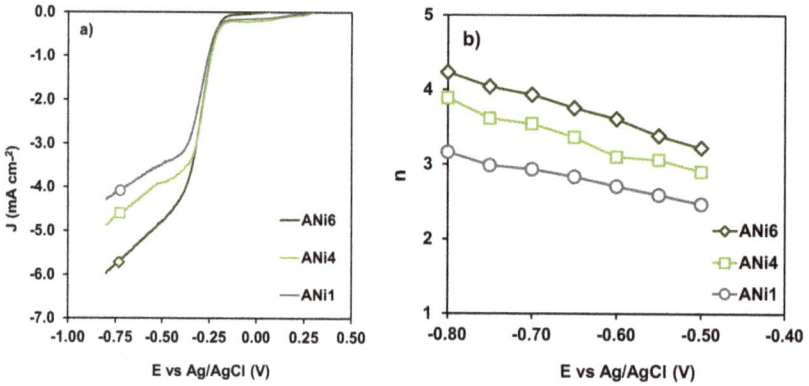

Figure 10. (**a**) LSV curves at 2500 rpm, and (**b**) variation of *n* with E vs. Ag/AgCl for samples ANi1 (○), ANi4 (□), ANi6 (◇).

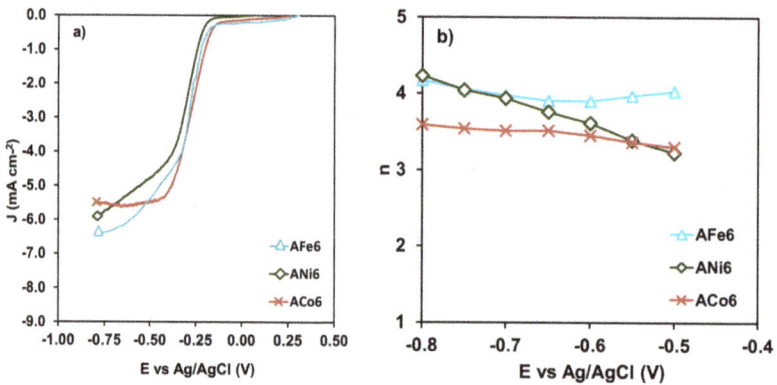

Figure 11. (**a**) LSV curves at 2500 rpm, and (**b**) variation of n with E vs. Ag/AgCl for samples AFe6 (△), ACo6 (×) and ANi6 (◇).

4. Discussion

After analyzing the textural data collected in Section 3, it is concluded that metal-doped aerogels contain a very well-developed porosity, especially with a significant mesoporosity (V_{BJH}). Aerogels doped with Ni have the highest surface areas and pore volumes, specially the highest micropore volumes. The metal phases are homogenously distributed into the carbon matrix and well dispersed. Doped aerogels contain a wide range of sizes of metal nano-particles, most of them with a zero-oxidation state (those embedded in the carbon matrix), with the exception of sample AFe6, which shows a mixture of Fe (0) and Fe (III). On the other hand, very low percentages of metal particles are detected in the external non-porous surface area, which would be partially oxidized. The macro-structure of these aerogels is similar among the samples; however, a partial graphitization process around the metal particles has also been detected in the case of the three different metals. In this line carbon aerogels doped with Ni seem to have the smallest and the best-developed graphitic clusters since their I_G/I_D values are the highest [35], which could be due to the fact that they have the smallest metal particles (Figure 4 and Table 2).

Regarding electro-catalytic experiments, it can be observed that increasing the Ni loading improved the electro-catalytic performance of the aerogel (Figure 10). In fact, when the Ni percentage was really small (ANi1), the oxygen reduction reaction occurs through a combination of the 2 and 4 e_s^-, denoted by a value of $n = 3.1$ (Table 3). Nevertheless, as the Ni content increased, the number of electrons transferred also did, and the oxygen reduction occurred with an electronic transfer of 4 e^-_s on both ANi4 and ANi6; although the reaction started at similar potentials as denoted by the value of E_{onset} (Table 3).

With respect to the type of metal, ACo6 is the best electro-catalyst showing the highest values of current density and the lowest value of E_{onset} among all the samples studied. On the other hand, some differences are also observed between ANi6 and AFe6 (Figure 11): ANi6 is the sample where the oxygen reduction occurs with a larger current density, although keeping the value of E_{onset} similar for both. According to bibliography [13,36,37], Ni-based electro-catalytic are in general less active than those-based in Fe or Co, this would be related to the ability of the metal to produce the dissociation of the oxygen molecule. However, in our material series, ANi6 show a very good electro-catalytic performance, even better than that for AFe6. In this case, ANi6 needs to be considered for its larger micropore volume and its smaller size of graphitic clusters. In fact, the electro-catalytic behavior of the carbon materials on the Oxygen Reduction Reaction is closely related to the type of carbon structure present in the material [38,39], and to its porosity. As was mentioned in previous paragraphs, the graphitic clusters for ANi6 are much smaller than those in the case of AFe6 and ACo6, therefore its good electro-catalytic performance could very well be related to it. In any case, it should be remarked that the catalytic results obtained with Ni-doped aerogels are especially interesting and much more in comparison to those obtained with Co- and Fe-doped ones.

On the other hand, we have included in Table 4 some bibliographic results obtained in similar experimental conditions to ours, and using as electro-catalysts Pt, Ni, Co, and Fe supported on different carbon materials. Pt/Carbon catalysts with a 20 wt % of Pt loading are a typical reference electrode [40–45], some authors use carbon black [40,43–45] and others prefer graphitic carbons as support [41,42]; in any case, all the collected results with this type of reference catalyst show the lowest E_{onset} potentials, which is reasonable because Pt itself is a better catalyst than the others, but the reported j_k values are lower or similar to ours. Despite that, our catalysts have much lower metal loadings and they do not contain platinum. Similar conclusions are obtained when carbon aerogels (prepared from melamine) [13,36], graphene oxide [46] or carbon nanotubes [44,45,47] were used as support of Ni, Co, or Fe; therefore, our j_k values are really good in comparison with those collected in Table 4. It is also remarkable that we have not found in the literature Ni-carbon-based electro-catalysts with a better performance than our ANi6 using similar ORR experimental conditions; its $j_k = 28.1$ is really significant. Thus, some electro-catalysts with low metal contents show n-values close to 2 which are lower than those obtained with our electro-catalysts ANi1 ($n = 3.1$). Finally, Ni and Co unsupported nanoparticles [48] have been described as not catalytically active in this experimental ORR condition. These results, together with the fact that our un-doped A0 aerogel neither was active in the reaction, would indicate some type of catalytic synergism effect between the carbon and metal phases, especially taking in account that the accessibility of the metal particles to the electrolyte could be limited by the graphitic cluster developed around them.

Table 4. Comparison of our electro-catalysts with others found in the literature using similar conditions and electrolyte KOH 0.1 M.

Catalyst Name	Type of Support	E_{onset} vs. Ag/AgCl (V)	n	Ref.	Metal wt %	j_k mA·cm^{-2}
ANi6	Carbon aerogel	−0.210	4.2	This work	5.8	28.1
ACo6	Carbon aerogel	−0.170	3.6	This work	5.8	34.9
AFe6	Carbon aerogel	−0.220	4.1	This work	6.1	26.2
20% Pt Vulcan	Carbon black	−0.037	3.9	[40]	20	N.R.
20% Pt/C	Graphitic carbon	−0.050	3.9	[41]	20	5
20% Pt/C	Graphitic carbon	−0.070	4.2	[42]	20	28.8
20% Pt/C	Carbon black	−0.065	4.0	[43]	20	14
20% Pt/C	Carbon black	-	3.9	[44]	20	≈29 *
Pt/Vulcan	Carbon black	−0.007	3.9	[45]	20	N.R.
NT_FePc_400	Carbon nanotube	−0.037	3.9	[45]	2.1	N.R.
NT_CoPc_400	Carbon nanotube	−0.150	2.4	[45]	2.1	N.R.
Co-NCA	Carbon aerogel	−0.150	4.0	[36]	3	≈25 *
Fe-NCA	Carbon aerogel	−0.150	3.8	[36]	5.2	≈14 *
Fe-NCA5	Carbon aerogel	−0.051	3.8	[13]	7.7	≈25 *
FeCo-N-rGO	Carbon nanotube	0.050	3.9	[44]	0.46	≈25 *
CoNPs/rGO	Graphene oxide	−0.115	3.9	[46]	0.3	N.R.
Fe3C-CNTFs	Carbon nanotube	0.105	3.1	[47]	N.R.	4.89
Co-CNTFs	Carbon nanotube	−0.015	3.9	[47]	N.R.	5.23
Ni-CNTs	Carbon nanotube	0.055	2.6	[47]	N.R.	3.67
Ni	Unsupported	0	0	[48]	100	0
Co	Unsupported	0	0	[48]	100	0

(N.R.) j_k values or j^{-1} vs. $\omega^{-1/2}$ plots not reported; (*) Data estimated from the corresponding j^{-1} vs. $\omega^{-1/2}$ plots.

Therefore, the results of this work clearly show that carbon aerogels doped with transition metals (obtained by polymerization of resorcinol and formaldehyde) are very good candidates as oxygen reduction electro-catalysts, where the current densities depend on the type and amount of metal doping and where the role of the carbon phase, both its textural and chemical properties, have a strong influence on the whole catalytic behavior of the material.

5. Conclusions

All the metal-doped carbon aerogels showed promising behavior in the oxygen reduction reaction; their well-developed porosity together with a very good metal dispersion in the carbon matrix, lead to materials with a very high electro-catalytic activity in ORR. As the Ni content was increased, the electro-catalytic behavior improved. Co-doped aerogel is the best electro-catalyst, showing the highest values of current density and the lowest value of E_{onset} among all the studied samples. Nevertheless, the nickel-doped aerogel (ANi6) presented even better results than the Fe one, which can be very well related to changes in the carbon crystalline structure and porosity, since ANi6 is the aerogel with the largest micropore and mesopore volumes and also the one with the smallest graphitic clusters. In general, the presence of small and well-developed graphitic domains seems to improve the electro-catalytic reduction of oxygen.

Acknowledgments: This research is supported by the FEDER and Spanish projects CTQ2013-44789-R (MINECO) and P12-RNM-2892 (Junta de Andalucía). A.A. is grateful to the European Union for his Erasmus Mundus fellowship, Program ELEMENT. J. C.-Q. is grateful to the Junta de Andalucía for her research contract (P12-RNM-2892). We thank the "Unidad de Excelencia Química Aplicada a Biomedicina y Medioambiente" (UGR) for its technical assistance".

Author Contributions: A.F.P.-C., M.P.-C. and F.C.-M. conceived and designed the experiments; A.A., J.C.-Q. and J.F.V.-V. performed the experiments; A.F.P.-C., F.C.-M, M.P.-C. and F.J.M.-H. analyzed the data; A.F.P.-C. and F.C.-M. wrote the paper.

Conflicts of Interest: The authors declare no conflict of interest.

References

1. Jung, D.H.; Bae, S.J.; Kim, S.J.; Nahm, K.S.; Kim, P. Effect of the Pt precursor on the morphology and catalytic performance of Pt-impregnated on Pd/C for the oxygen reduction reaction in polymer electrolyte fuel cells. *Int. J. Hydrog. Energy* **2011**, *36*, 9115–9122. [CrossRef]

2. Yang, Z.; Nie, H.; Chen, X.; Chen, X.; Huang, S. Recent progress in doped carbon nanomaterials as effective cathode catalysts for fuel cell oxygen reduction reaction. *J. Power Sources* **2013**, *236*, 238–249. [CrossRef]

3. Banham, D.; Ye, S.; Pei, K.; Ozaki, J.-I.; Kishimoto, T.; Imashiro, Y. A review of the stability and durability of non-precious metal catalysts for the oxygen reduction reaction in proton exchange membrane fuel cells. *J. Power Sources* **2015**, *285*, 334–348. [CrossRef]

4. Lee, C.L.; Chiou, H.P.; Syu, C.M.; Wu, C.C. Silver triangular nanoplates as electrocatalyst for oxygen reduction reaction. *Electrochem. Commun.* **2010**, *12*, 1609–1613. [CrossRef]

5. Bo, X.; Zhang, Y.; Li, M.; Nsabimana, A.; Guo, L. NiCo$_2$O$_4$ spinel/ordered mesoporous carbons as noble-metal free electrocatalysts for oxygen reduction reaction and the influence of structure of catalyst support on the electrochemical activity of NiCo$_2$O$_4$. *J. Power Sources* **2015**, *288*, 1–8. [CrossRef]

6. Stojmenović, M.; Momčilović, M.; Gavrilov, N.; Pašti, I.A.; Mentus, S.; Jokić, B.; Babić, B. Incorporation of Pt, Ru and Pt-Ru nanoparticles into ordered mesoporous carbons for efficient oxygen reduction reaction in alkaline media. *Electrochim. Acta* **2015**, *153*, 130–139. [CrossRef]

7. Zhao, A.; Masa, J.; Xia, W. Oxygen-deficient titania as alternative support for Pt catalysts for the oxygen reduction reaction. *J. Energy Chem.* **2014**, *23*, 701–707. [CrossRef]

8. Kim, J.H.; Chang, S.; Kim, Y.T. Compressive strain as the main origin of enhanced oxygen reduction reaction activity for Pt electrocatalysts on chromium-doped titania support. *Appl. Catal. B Environ.* **2014**, *158–159*, 112–118. [CrossRef]

9. Ahmed, M.S.; Kim, D.; Jeon, S. Covalently grafted platinum nanoparticles to multi walled carbon nanotubes for enhanced electrocatalytic oxygen reduction. *Electrochim. Acta* **2013**, *92*, 168–175. [CrossRef]

10. Oh, J.-M.; Park, J.; Kumbhar, A.; Smith, D.; Creager, S. Electrochemical Oxygen Reduction at Platinum/Mesoporous Carbon/Zirconia/Ionomer Thin-Film Composite Electrodes. *Electrochim. Acta* **2014**, *138*, 278–287. [CrossRef]

11. Dou, S.; Shen, A.; Ma, Z.; Wu, J.; Tao, L.; Wang, S. N-, P- and S-tridoped graphene as metal-free electrocatalyst for oxygen reduction reaction. *J. Electroanal. Chem.* **2015**, *753*, 21–27. [CrossRef]

12. Ishii, T.; Maie, T.; Kimura, N.; Kobori, Y.; Imashiro, Y.; Ozaki, J.-I. Enhanced catalytic activity of nanoshell carbon co-doped with boron and nitrogen in the oxygen reduction reaction. *Int. J. Hydrog. Energy* **2017**, *42*, 5–12. [CrossRef]

13. Sarapuu, A.; Kreek, K.; Kisand, K.; Kook, M.; Uibu, M.; Koel, M.; Tammeveski, K. Electrocatalysis of oxygen reduction by iron-containing nitrogen-doped carbon aerogels in alkaline solution. *Electrochim. Acta* **2017**, *230*, 81–88. [CrossRef]

14. Elmouwahidi, A.; Vivo-Vilches, J.F.; Pérez-Cadenas, A.F.; Maldonado-Hódar, F.J.; Carrasco-Marín, F. Free metal oxygen-reduction electro-catalysts obtained from biomass residue of the olive oil industry. *Chem. Eng. J.* **2016**, *306*, 1109–1115. [CrossRef]

15. Pekala, R.W.; Alviso, C.T.; Kong, F.M.; Hulsey, S.S. Aerogels derived from multifunctional organic monomers. *J. Non-Cryst. Solids* **1992**, *145*, 90–98. [CrossRef]

16. ElKhatat, A.M.; Al-Muhtaseb, S.A. Advances in tailoring resorcinol-formaldehyde organic and carbon gels. *Adv. Mater.* **2011**, *23*, 2887–2903. [CrossRef] [PubMed]

17. Gallegos-Suárez, E.; Pérez-Cadenas, A.F.; Maldonado-Hódar, F.J.; Carrasco-Marín, F. On the micro-and mesoporosity of carbon aerogels and xerogels. The role of the drying conditions during the synthesis processes. *Chem. Eng. J.* **2012**, *181*, 851–855. [CrossRef]

18. Morales-Torres, S.; Maldonado-Hódar, F.J.; Pérez-Cadenas, A.F.; Carrasco-Marín, F. Textural and mechanical characteristics of carbon aerogels synthesized by polymerization of resorcinol and formaldehyde using alkali carbonates as basification agents. *Phys. Chem. Chem. Phys.* **2010**, *12*, 10365–10372. [CrossRef] [PubMed]
19. Maldonado-Hódar, F.J.; Jirglová, H.; Pérez-Cadenas, A.F.; Morales-Torres, S. Chemical control of the characteristics of Mo-doped carbon xerogels by surfactant-mediated synthesis. *Carbon N. Y.* **2013**, *51*, 213–223. [CrossRef]
20. Vivo-Vilches, J.F.; Carrasco-Marín, F.; Pérez-Cadenas, A.F.; Maldonado-Hódar, F.J. Fitting the porosity of carbon xerogel by CO_2 activation to improve the TMP/n-octane separation. *Microporous Mesoporous Mater.* **2015**, *209*, 10–17. [CrossRef]
21. Maldonado-Hódar, F.J.; Moreno-Castilla, C.; Carrasco-Marín, F.; Pérez-Cadenas, A.F. Reversible toluene adsorption on monolithic carbon aerogels. *J. Hazard. Mater.* **2007**, *148*, 548–552. [CrossRef] [PubMed]
22. Bailon-Garcia, E.; Carrasco-Marin, F.; Perez-Cadenas, A.F.; Maldonado-Hodar, F.J. Microspheres of carbon xerogel: An alternative Pt-support for the selective hydrogenation of citral. *Appl. Catal. A Gen.* **2014**, *482*, 318–326. [CrossRef]
23. Bailón-García, E.; Elmouwahidi, A.; Álvarez, M.A.; Carrasco-Marín, F.; Pérez-Cadenas, A.F.; Maldonado-Hódar, F.J. New carbon xerogel-TiO_2 composites with high performance as visible-light photocatalysts for dye mineralization. *Appl. Catal. B Environ.* **2017**, *201*, 29–40. [CrossRef]
24. Duarte, F.; Maldonado-Hódar, F.J.; Pérez-Cadenas, A.F.; Madeira, L.M. Fenton-like degradation of azo-dye Orange II catalyzed by transition metals on carbon aerogels. *Appl. Catal. B Environ.* **2009**, *85*, 139–147. [CrossRef]
25. Maldonado-Hódar, F.J.; Moreno-Castilla, C.; Pérez-Cadenas, A.F. Catalytic combustion of toluene on platinum-containing monolithic carbon aerogels. *Appl. Catal. B Environ.* **2004**, *54*, 217–224. [CrossRef]
26. Elmouwahidi, A.; Bailón-García, E.; Pérez-Cadenas, A.F.; Maldonado-Hódar, F.J.; Castelo-Quibén, J.; Carrasco-Marín, F. Electrochemical performances of supercapacitors from carbon-ZrO_2 composites. *Electrochim. Acta* **2018**, *259*, 803–814. [CrossRef]
27. Elmouwahidi, A.; Bailón-García, E.; Castelo-Quibén, J.; Pérez-Cadenas, A.F.; Maldonado-Hódar, F.J.; Carrasco-Marín, F. Carbon-TiO_2 composites as high-performance supercapacitor electrodes: Synergistic effect between carbon and metal oxide phases. *J. Mater. Chem. A* **2018**, *6*. [CrossRef]
28. Pérez-Cadenas, A.F.; Maldonado-Hódar, F.J.; Moreno-Castilla, C. Molybdenum carbide formation in molybdenum-doped organic and carbon aerogels. *Langmuir* **2005**, *21*, 10850–10855. [CrossRef] [PubMed]
29. Maldonado-Hódar, F.J.; Pérez-Cadenas, A.F.; Moreno-Castilla, C. Morphology of heat-treated tunsgten doped monolithic carbon aerogels. *Carbon N. Y.* **2003**, *41*, 1291–1299. [CrossRef]
30. Barrett, E.P.; Joyner, L.G.; Halenda, P.P. The determination of pore volume and area distributions in porous substances. I. Computations from nitrogen isotherms. *J. Am. Chem. Soc.* **1951**, *73*, 373–380. [CrossRef]
31. Maldonado-Hódar, F.J.; Moreno-Castilla, C.; Pérez-Cadenas, A.F. Surface morphology, metal dispersion, and pore texture of transition metal-doped monolithic carbon aerogels and steam-activated derivatives. *Microporous Mesoporous Mater.* **2004**, *69*, 119–125. [CrossRef]
32. Maldonado-Hódar, F.J.; Moreno-Castilla, C.; Rivera-Utrilla, J.; Hanzawa, Y.; Yamada, Y. Catalytic Graphitization of Carbon Aerogels by Transition Metals. *Langmuir* **2000**, *16*, 4367–4373. [CrossRef]
33. NIST X-ray Photoelectron Spectroscopy Database. Available online: https://srdata.nist.gov/xps/ (accessed on 16 April 2018).
34. Schwan, J.; Ulrich, S.; Batori, V.; Ehrhardt, H.; Silva, S.R.P. Raman spectroscopy on amorphous carbon films. *J. Appl. Phys.* **1996**, *80*, 440. [CrossRef]
35. Wang, S.; Xu, Y.; Yan, M.; Zhai, Z.; Ren, B.; Zhang, L.; Liu, Z. Comparative study of metal-doped carbon aerogel: Physical properties and electrochemical performance. *J. Electroanal. Chem.* **2018**, *809*, 111–116. [CrossRef]
36. Sarapuu, A.; Samolberg, L.; Kreek, K.; Koel, M.; Matisen, L.; Tammeveski, K. Cobalt- and iron-containing nitrogen-doped carbon aerogels as non-precious metal catalysts for electrochemical reduction of oxygen. *J. Electroanal. Chem.* **2015**, *746*, 9–17. [CrossRef]
37. Chen, Z.; Higgins, D.; Yu, A.; Zhang, L.; Zhang, J. A review on non-precious metal electrocatalysts for PEM fuel cells. *Energy Environ. Sci.* **2011**, *4*, 3167–3192. [CrossRef]
38. Shin, D.; An, X.; Choun, M.; Lee, J. Effect of transition metal induced pore structure on oxygen reduction reaction of electrospun fibrous carbon. *Catal. Today* **2016**, *260*, 82–88. [CrossRef]

39. Chao, S.; Zhang, Y.; Wang, K.; Bai, Z.; Yang, L. Flower—like Ni and N codoped hierarchical porous carbon microspheres with enhanced performance for fuel cell storage. *Appl. Energy* **2016**, *175*, 421–428. [CrossRef]

40. Quílez-Bermejo, J.; González-Gaitán, C.; Morallón, E.; Cazorla-Amorós, D. Effect of carbonization conditions of polyaniline on its catalytic activity towards ORR. Some insights about the nature of the active sites. *Carbon N. Y.* **2017**, *119*, 62–71. [CrossRef]

41. Ferrero, G.A.; Preuss, K.; Fuertes, A.B.; Sevilla, M.; Titirici, M.-M. The influence of pore size distribution on the oxygen reduction reaction performance in nitrogen doped carbon microspheres. *J. Mater. Chem. A* **2016**, *4*, 2581–2589. [CrossRef]

42. Zhao, X.; Zhao, H.; Zhang, T.; Yan, X.; Yuan, Y.; Zhang, H.; Zhao, H.; Zhang, D.; Zhu, G.; Yao, X. One-step synthesis of nitrogen-doped microporous carbon materials as metal-free electrocatalysts for oxygen reduction reaction. *J. Mater. Chem. A* **2014**, *2*, 11666–11671. [CrossRef]

43. Pan, F.; Cao, Z.; Zhao, Q.; Liang, H.; Zhang, J. Nitrogen-doped porous carbon nanosheets made from biomass as highly active electrocatalyst for oxygen reduction reaction. *J. Power Sources* **2014**, *272*, 8–15. [CrossRef]

44. Fu, X.; Liu, Y.; Cao, X.; Jin, J.; Liu, Q.; Zhang, J. FeCo–Nx embedded graphene as high performance catalysts for oxygen reduction reaction. *Appl. Catal. B Environ.* **2013**, *130–131*, 143–151. [CrossRef]

45. González-Gaitán, C.; Ruiz-Rosas, R.; Morallón, E.; Cazorla-Amorós, D. Relevance of the Interaction between the M-Phthalocyanines and Carbon Nanotubes in the Electroactivity toward ORR. *Langmuir* **2017**, *33*, 11945–11955. [CrossRef] [PubMed]

46. Liu, X.; Yu, Y.; Niu, Y.; Bao, S.; Hu, W. Cobalt nanoparticle decorated graphene aerogel for efficient oxygen reduction reaction electrocatalysis. *Int. J. Hydrog. Energy* **2017**, *42*, 5930–5937. [CrossRef]

47. Wang, M.-Q.; Ye, C.; Wang, M.; Li, T.-H.; Yu, Y.-N.; Bao, S.-J. Synthesis of M (Fe₃C, Co, Ni)-porous carbon frameworks as high-efficient ORR catalysts. *Energy Storage Mater.* **2018**, *11*, 112–117. [CrossRef]

48. Goubert-Renaudin, S.N.S.; Wieckowski, A. Ni and/or Co nanoparticles as catalysts for oxygen reduction reaction (ORR) at room temperature. *J. Electroanal. Chem.* **2011**, *652*, 44–51. [CrossRef]

nanomaterials

MDPI

Article

Tailoring of Perpendicular Magnetic Anisotropy in Dy$_{13}$Fe$_{87}$ Thin Films with Hexagonal Antidot Lattice Nanostructure

Mohamed Salaheldeen [1,2], **Victor Vega** [2,3], **Angel Ibabe** [4], **Miriam Jaafar** [4], **Agustina Asenjo** [4], **Agustin Fernandez** [2] **and Victor M. Prida** [2,*]

1   Physics Department, Faculty of Science, Sohag University, 82524 Sohag, Egypt; UO253675@uniovi.es
2   Depto. Física, Universidad de Oviedo, C/Federico Garcia Lorca 18, 33007 Oviedo, Asturias, Spain; vegavictor@uniovi.es (V.V.); aafernandez@uniovi.es (A.F.)
3   Laboratorio Membranas Nanoporosas, Servicios Científico-Técnicos, Universidad de Oviedo, Campus El Cristo s/n, 33006 Oviedo, Asturias, Spain
4   Instituto de Ciencia de Materiales de Madrid, CSIC, Cantoblanco, 28049 Madrid, Spain; angel.ibabe@estudiante.uam.es (A.I.); m_jaafar@icmm.csic.es (M.J.); aasenjo@icmm.csic.es (A.A.)
*   Correspondence: vmpp@uniovi.es; Tel.: +34-985-103-294

Received: 10 March 2018; Accepted: 5 April 2018; Published: 8 April 2018

Abstract: In this article, the magnetic properties of hexagonally ordered antidot arrays made of Dy$_{13}$Fe$_{87}$ alloy are studied and compared with corresponding ones of continuous thin films with the same compositions and thicknesses, varying between 20 nm and 50 nm. Both samples, the continuous thin films and antidot arrays, were prepared by high vacuum e-beam evaporation of the alloy on the top-surface of glass and hexagonally self-ordered nanoporous alumina templates, which serve as substrates, respectively. By using a highly sensitive magneto-optical Kerr effect (MOKE) and vibrating sample magnetometer (VSM) measurements an interesting phenomenon has been observed, consisting in the easy magnetization axis transfer from a purely in-plane (INP) magnetic anisotropy to out-of-plane (OOP) magnetization. For the 30 nm film thickness we have measured the volume hysteresis loops by VSM with the easy magnetization axis lying along the OOP direction. Using magnetic force microscopy measurements (MFM), there is strong evidence to suggest that the formation of magnetic domains with OOP magnetization occurs in this sample. This phenomenon can be of high interest for the development of novel magnetic and magneto-optic perpendicular recording patterned media based on template-assisted deposition techniques.

Keywords: nanoporous alumina templates; antidot arrays; Kerr effect; magnetic anisotropy; magnetic domains; magnetic force microscopy

1. Introduction

The engineering of magnetic systems based on planar patterned nanostructures by properly controlling the shape and size of the patterned nano-objects allows obtaining magnetic films with non-collinear magnetization distribution, thus enabling the development of magnetic data processing devices with vertical architecture and spin-based electronics [1]. Nanoscale antidot arrays lattices in magnetic materials, i.e., periodic spatial arrangements of nanometric holes in thin metal films, have been studied in a wide scientific and technical scope. Tuned magnetic frustration in spin ice [2,3] and spin glass [4] behavior have been investigated and detected in magnetic nanoscale antidot arrays. The extraordinary features exhibited by these nanostructured materials comes from their relatively large surface to volume ratio, which influences the shape, morphology, crystalline structure, and surface roughness of the material in the nanoscale range [5,6]. These artificial arrays have been recently used in ultra-high density data storage applications [7], where the antidot lattices allow to define a peculiar type of bit patterned media that can overcome the superparamagnetic limit due to the non-formation of isolated magnetic domains [8], in addition to magnetic bio-sensing applications [9], among many others. Moreover, they attract notable attention due to their capacity to act as metamaterials (magnonic crystals), where the antidots exhibit a periodic potential for magnons, allowing for the control of spin wave dispersion [10,11]. The recently reported nanostructured material based on the exchange coupled bicomposite formed by Co. dots embedded in a matrix of NiFe antidot arrays, where the soft ferromagnetic properties of the NiFe antidots influence the vortex nucleation field of Co. dots, can modify the magnetotransport and spin wave properties of the system [12]. Therefore, a wide range of applications in the field of spintronics and spin wave filtering are available by using antidots arrays. The competition between the intrinsic thin film and local shape anisotropies, together with the local effects created by the antidots arrays, generates a new scenario for tailoring the magnetic properties of the thin films by suitably tuning their geometric parameters [13,14]. Furthermore, antidot lattices strongly influence the magnetic properties of the hosting materials and can be used for artificially engineering the magnetic anisotropy and tuning the coercivity in thin films [15]. In this work, we report on the observation of an interesting phenomenon, consisting of the easy magnetization axis transfer from a purely in-plane (INP) magnetic anisotropy to out-of-plane (OOP) magnetization, for a 30 nm thick layer of $Dy_{13}Fe_{87}$ thin film with hexagonally ordered antidots lattice.

2. Materials and Methods

Experimental Procedure of Sample Fabrication and Characterization

The former patterned film substrate formed by a hexagonally ordered nanoporous alumina membrane was synthesized by following a well-established two-step electrochemical anodization procedure in oxalic acid, as reported elsewhere [16,17]. In brief, 0.5 mm thick, high purity Al foils (99.999%, Goodfellow, Huntingdon, UK) were cleaned by sonication in ethanol and isopropyl alcohols and electropolished at 20 V in perchloric acid and ethanol solution (1:3 vol., 5 °C) for 5 min. The polished Al foils were then employed as starting substrates for the anodic synthesis of nanoporous alumina templates. The two-step electrochemical anodization was performed in 0.3 M oxalic acid, at a temperature of 1–3 °C and under a potentiostatic applied voltage of 40 V, measured versus a Pt counter electrode. Between the two anodization steps, the samples were immersed in 0.2 M CrO_3 and 0.6 M H_3PO_4 aqueous solution. This selective chemical etching step allowed for the selective removal of the first grown anodic alumina layer, which contained randomly disordered nanopores at its top surface. In the second anodization step, which lasted for 5 h, the nanopores grew following a highly self-ordered hexagonal pattern. Afterwards, in order to increase the pore size, the samples were chemically etched in 5 wt % orthophosphoric acid at 30 °C, for 30 min.

The controlled deposition of the metallic alloy formed by highly pure metal pieces of Dy (99.99%) and Fe (99.9%) was completed by a high vacuum thermal evaporation technique using an E306A thermal vacuum coating unit (Edwards, Crawley, UK) with an ultimate vacuum better than 7×10^{-7} mbar (5.2×10^{-7} mbar), having a diffusion pump backed by rotary pumping together with a liquid nitrogen trap [18]. The pure element metal pieces were placed inside two water cooled copper crucibles and heated by the action of magnetically focused electron beams. The evaporated target metals were deposited on the top-surface of the hexagonally ordered nanoporous alumina membranes, which acted as templates to obtain the thin film antidot arrays [19]. The control of the film thickness and the alloy composition was achieved by using two independent quartz crystal controllers that monitored simultaneously the deposition rates of both evaporation sources. This equipment allowed for obtaining both the film thickness and final alloy composition from the measurements displayed in both of the quartz crystal control monitors. Each one of these quartz crystal controllers received the evaporation beam coming from a unique evaporation source. The source to substrate distances were maintained constant at about 18 cm. The deposition rate was around 0.1–0.15 nm s^{-1}.

The alloy composition was controlled through the quartz crystal control, and the values obtained by this procedure are in good agreement with the ones analyzed by energy dispersive X-ray spectroscopy (EDX) (Inca Energy 200, Oxford Instruments, Abingdon, UK) with a scanning electron microscope (SEM) (JSM 5600, JEOL, Akishima, Tokyo, Japan).

High resolution transmission electron microscopy (HR-TEM) (JEM 2100, JEOL, Akishima, Tokyo, Japan) operating at 200 kV was employed to obtain high magnification images of the antidots thin films, as well as to study their microstructure by performing a selected area electron diffraction (SAED) spectra. For that purpose, the nanoporous alumina membrane that acts as a template for the fabrication of the antidot arrays was previously and selectively dissolved in a 0.5 M NaOH solution, thus releasing freestanding flakes of the nanostructured thin film, which were then washed with distilled water and ethanol, deposited into conventional transmission electron microscopy (TEM) copper grid sample holders and dried in air.

The surface topography and magnetic domain configuration were studied by Atomic Force Microscopy (AFM) and Magnetic Force Microscopy (MFM) measurements, respectively, performed with a Cervantes system from Nanotec Electronica S.L. (Tres Cantos, Madrid, Spain) in amplitude modulation mode with the Phase Locked Loop (PLL) feedback enabled. Commercial probes from Budget Sensors MagneticMulti75-G, with CoCr coating were used.

The surface magneto-optic properties of the thin film antidot arrays were obtained making use of a scanning laser Magneto-Optical Kerr Effect (MOKE) magnetometer, NanoMOKE3® (Durham Magneto Optics Ltd., Durham, UK), being able to apply up to 0.125 T by using the quadrupole electromagnet option or 0.5 T, with the dipole electromagnet option. The NanoMOKE3 magnetometer is suited with p-polarized laser beam and it is sensitive to the longitudinal, transversal, and polar magneto-optical Kerr effects. Complementary bulk magnetic measurements were carried out by using a vibrating sample magnetometer (VSM-QD-Versalab, San Diego, CA, USA), with applied magnetic fields up to ±3 T at room temperature and in both, parallel (In Plane, INP) and perpendicular (Out of Plane, OOP) directions to the film plane, respectively.

3. Results and Discussion

3.1. Scaning Electron Microscopy Analysis

Figure 1a displays the highly ordered, hexagonally centered nanopores of the alumina template after being synthesized by two-step electrochemical anodization in oxalic acid and further pore widening by chemical etching, as explained in detail in the experimental section. The resulting lattice parameters of the so-synthesized nanoporous alumina membrane employed as a starting template for the thin film deposition are around 75 nm of nanopores diameter, dp, and 105 nm of the interpore distance, $Dint$. Hexagonal-ordered array of $Dy_{13}Fe_{87}$ antidot thin films having an interpore distance of $Dint = 105$ nm, hole diameter $d = 45$ nm, and thickness $t = 30$ nm, are shown in Figure 1b. The antidot thin film displayed in Figure 1b replicates the same hexagonal nanoholes ordering of the starting nanoporous alumina membrane used as a patterned template, as shown in Figure 1a. Therefore, the antidot array structure exhibits two main distinguished directions [13]: first, the nearest neighbors (*nn*) direction, which corresponds to the in-plane easy anisotropy axis; and second, the next-nearest neighbors (*nnn*) direction along which lies the in-plane hard anisotropy axis, at an angle of 30° referred to the previous *nn* direction (see Figure 1b). Figure 1c shows a 50 nm thick antidot thin film, deposited onto the surface of the patterned nanoporous alumina template as shown in Figure 1a. It becomes evident from the comparison of both SEM images displayed in Figure 1a–c that there is a nearly inversed linear dependence between the diameter of the antidots, d, and the film thickness, t, for a given pore diameter of the nanoporous alumina template, dp. In other words, as the thin film thickness increases, the nanoholes decrease their manifest diameter due to the deposition of hosting material on the top part of the hole wall [20,21].

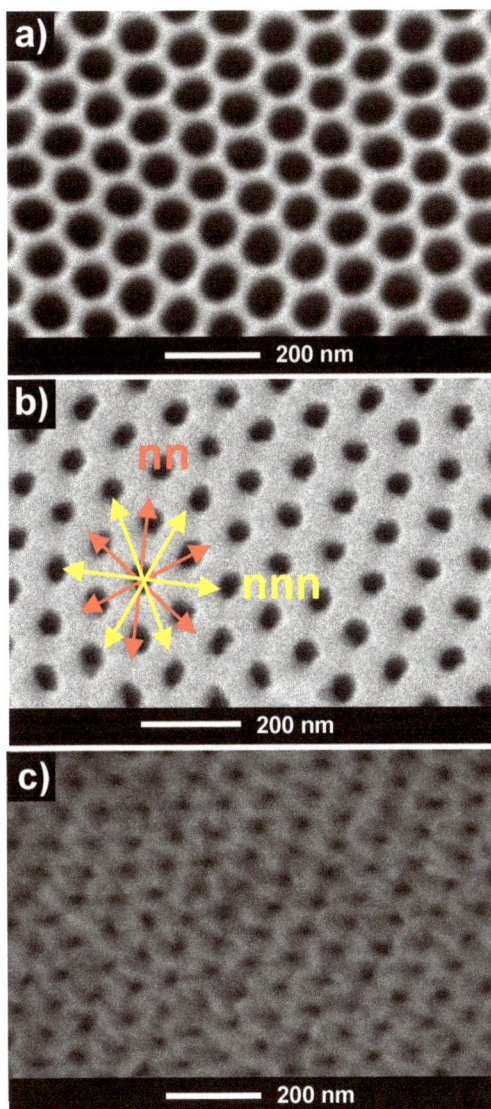

Figure 1. (a) Top-view surface Scanning Electron Microscope (SEM) image of a nanoporous alumina membrane employed as patterned substrate for depositing the antidots arrays showing the hexagonal ordering of the nanopores, (interpore distance *Dint* = 105 nm, pore diameter *dp* = 75 nm); (b) SEM top-view image of the surface of 30 nm thick $Dy_{13}Fe_{87}$ thin film antidot array, obtained by replicating the alumina template shown in (a). The in-plane easy anisotropy axis, *nn*, and in-plane hard anisotropy direction, *nnn*, are indicated by red and yellow arrows, respectively; (c) SEM top-view of the antidot array $Dy_{13}Fe_{87}$ sample with 50 nm in thickness.

3.2. Transmission Electron Microscopy Characterization

The TEM micrograph of the $Dy_{13}Fe_{87}$ antidots thin film after being released from the nanoporous alumina membrane, which is displayed in Figure 2, demonstrates that the nanometric holes successfully replicated the structure of the highly hexagonal ordered nanoporous alumina template, in good agreement with the findings revealed by the SEM images. The Fast Fourier Transform (FFT) shown in the upper inset of Figure 2, which is characteristic of a system with hexagonal periodic symmetry, reinforces this observation. In addition, the SAED spectrum, shown as the lower inset in Figure 2, indicates the amorphous structure of the $Dy_{13}Fe_{87}$ alloy, evidenced by the presence of diffused rings and the absence of clear spots in the electron diffraction spectrum.

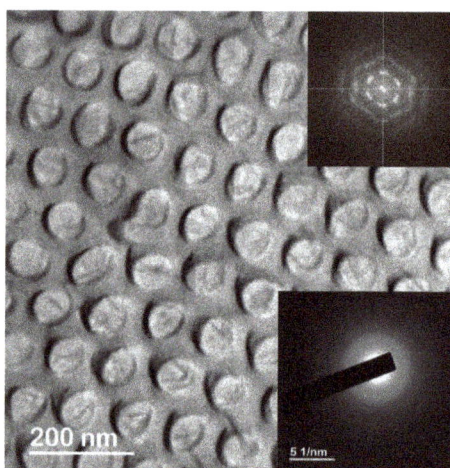

Figure 2. Transmission Electron Microscopy (TEM) image of a small region of $Dy_{13}Fe_{87}$ antidots thin film after being released from the nanoporous alumina template. The Fast Fourier Transform (FFT) analysis shown in the upper inset reveals the highly hexagonal ordering degree of the antidots by replicating the nanoporous structure of the patterned alumina membrane, while the electron diffraction pattern displayed in the lower inset demonstrates the amorphous character of the deposited $Dy_{13}Fe_{87}$ alloy.

3.3. Atomic Force Microscopy and Magnetic Force Microscopy Imaging

The AFM image of the surface topography of the $Dy_{13}Fe_{87}$ antidot sample, with thickness of 30 nm, can be observed in Figure 3a. The nanometric hole structure of the magnetic film is clearly visible and replicates the hexagonal geometry of the nanoporous alumina membrane used as template. The MFM signal is mainly sensitive to the out of plane component of the magnetization and to the magnetic-poles accumulation around the domain walls. The MFM image obtained in the same region in the *as prepared* state (see Figure 3b) presents a magnetic configuration with positive and negative contrast corresponding to the out of plane magnetization. As shown in Figure 3a, the sample presents different morphological domains. Figure 3c,d are the zooms performed in the marked regions in Figure 3a,b, respectively, in order to study the topographic periodicity and its correlation with the magnetic signal. The profiles along the *nn* direction are displayed in Figure 3e,f. From the profile scan in Figure 3e, as well as from the FFT data shown in the inset, the lattice constant of the hexagonal antidot arrangement is estimated to be around 110 nm, in good agreement with SEM and TEM characterization. The corresponding MFM signal in Figure 3d,f allows us to distinguish the superposition of two kinds of contrast, one correlated with the topographic signal (always negative) and a positive/negative contrast corresponding to the out of plane magnetization, which is not correlated with the holes' position. Notice that the FFT of the MFM signal presents a set of peaks corresponding to the hexagonal lattice and an additional contrast associated to the larger structures. Such component of the magnetic signal is due to the local magnetic anisotropy induced by the antidot thin film geometry.

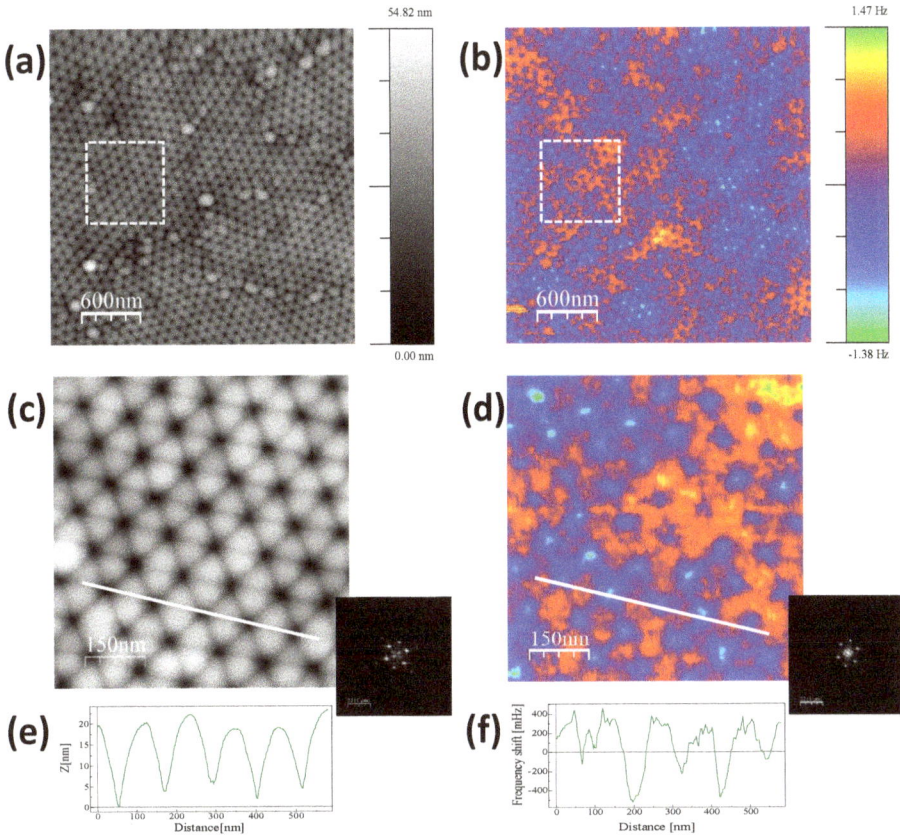

Figure 3. (**a**) Topography and (**b**) Magnetic Force Microscopy (MFM) images corresponding to the $Dy_{13}Fe_{87}$ antidot sample with 30 nm in thickness; (**c**,**d**) are zooms corresponding to the marked region in (**a**,**b**), respectively. Insets correspond to the FFT of each image; (**e**,**f**) are the profiles obtained along the marked lines in (**c**,**d**). The magnetic state of the sample is *as prepared*.

3.4. Magneto-Optical Kerr Effect Hysteresis Loops

The surface magnetic properties of the $Dy_{13}Fe_{87}$ antidot thin films were characterized making use of the MOKE. In the Figure 4a it is represented, in black line, the longitudinal MOKE hysteresis loop of a continuous $Dy_{13}Fe_{87}$ thin film, with 30 nm in thickness. It shows a square shape that matches with an in-plane easy magnetization axis. Red and blue plots in Figure 4a show the longitudinal MOKE hysteresis loops for 30 nm thick antidot thin film, measured along both, the *nn* and *nnn* directions, respectively. By comparison of these two later measurements, we can observe that these directions act as easy (*nn*) and hard (*nnn*) in-plane magnetization axes. Furthermore, both longitudinal hysteresis loops for the antidot thin film measured along the in-plane *nn* and *nnn* directions have lost the squared shape and acquired a shape bending in slope, if compared with the one from the continuous thin film. This fact shows that the magnetization is not aligned along an easy axis direction, indicating that the easy magnetization axis does not totally lie in the in-plane direction. Coercivity of $Dy_{13}Fe_{87}$ antidots thin film also noticeably increases with respect to the one for the continuous film, caused by the presence of holes that act as pinning centers of domain wall displacement [15].

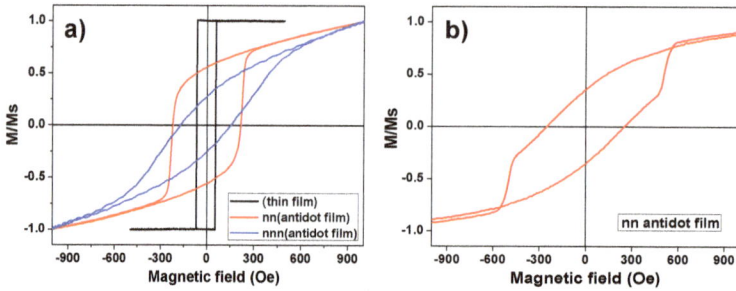

Figure 4. (**a**): Longitudinal Magneto-Optical Kerr Effect (MOKE) hysteresis loops of a 30 nm thick $Dy_{13}Fe_{87}$ thin film (black line) and antidot film of the same thickness and composition, measured along the *nn* (red line) and *nnn* (blue line) directions of the hexagonal lattice. (**b**): Polar MOKE hysteresis loop with an out of plane applied magnetic field for the same antidot thin film.

Figure 4b shows the hysteresis loop obtained by polar MOKE for the same antidot thin film with an out-of-plane applied magnetic field. In the polar MOKE configuration, the response of the polar Kerr effect is sensitive to the out of plane magnetization signal, while the longitudinal Kerr measurement is sensitive to any magnetization along the intersection of the film surface plane with the incidence plane of the laser beam, and thus the longitudinal Kerr response is dominated by the in-plane magnetization [22]. The hysteresis loop plotted in Figure 4b shows a dominant magnetization component perpendicular to the surface plane in pseudo magnetization saturation. This fact proves the dropping of the magnetization onto the out of plane direction [22].

3.5. Vibrating Sample Magnetometer Hysteresis Loops

The global magnetic behavior of antidots thin film has been investigated by measuring the bulk sample hysteresis loops, employing a VSM, at room temperature and with applied magnetic field values up to ±3 T. Two common magnetic field configurations were considered for the applied magnetic field, in-plane (INP, when the magnetic field is applied parallel to the film plane) and out of plane (OOP, for the magnetic field perpendicularly applied to film plane). Figure 5a shows both, the INP and OOP hysteresis loops for a continuous thin film having 30 nm in thickness and Figure 5b shows the corresponding ones for the antidot thin film with the same thickness.

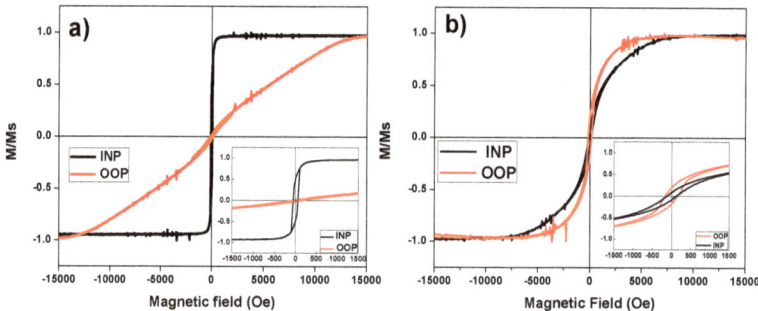

Figure 5. (**a**) Vibrating Sample Magnetometer In Plane (VSM INP) and Out Of Plane (OOP) hysteresis loops for the continuous $Dy_{13}Fe_{87}$ thin film with 30 nm thickness. The inset shows the low field scale of INP and OOP hysteresis loops; (**b**) INP and OOP VSM hysteresis loops for $Dy_{13}Fe_{87}$ antidot film with 30 nm thickness, *Dint* = 105 nm and *d* = 45 nm. The inset shows the magnification of the low field region.

It can be observed from Figure 5a that the magnetization is laying along the INP direction for the $Dy_{13}Fe_{87}$ continuous film. In contrast, the OOP magnetization direction for antidot arrays is dominating, as it can be deduced from Figure 5b and confirmed by the polar-MOKE measurement. In addition, it can be observed the INP shape anisotropy deduced for the continuous thin film, while a more complex magnetic structure is ascribed for the antidot thin film, where its OOP component clearly differs from that of the continuous film. Similar magnetic behavior was also detected by J. Gräfe et al. in Fe antidot arrays with hexagonal ordering [23].

It is expected that the magnetic anisotropy contributes to the appearance of the OOP magnetization component. The antidot arrays can induce strong local shape anisotropy. This can overcome any kind of intrinsic anisotropy of the host materials, moreover it prefers an OOP orientation of the magnetization [24]. In addition, theoretical studies about the magnetic anisotropy of antidot arrays performed by Monte Carlo [25] and in micro-magnetic simulations [26] illustrate that the INP preferred orientation of the magnetization in a thin film of antidots array can be, at least partially, lifted. Furthermore, the magnetic surface anisotropy contributes to the partial OOP magnetization found here [27].

The local geometry of antidot plays a major important role on the magnetization reversal of the antidot films [28–31]. Such kind of nanohole arrays are usually described by a parameter named the antidot aspect ratio, r, which is defined by [32,33]:

$$r = (d + dp)/2t,$$ (1)

Aiming to further investigate the rise of OOP magnetization phenomena, we have estimated the effective in-plane anisotropy, K_{eff}, for antidots and thin film samples with the same $Dy_{13}Fe_{87}$ composition and different thicknesses. K_{eff} values were calculated from INP and OOP VSM hysteresis loops according to ref. [29], and they are given by:

$$K_{eff} = 4\pi \left[\int_{0\_OOP}^{M_S} H dM - \int_{0\_INP}^{M_S} H dM \right]$$ (2)

where M is the magnetization, M_S is the saturation magnetization and H is the applied magnetic field. These effective anisotropy values obtained for samples with different thickness of the $Dy_{13}Fe_{87}$ film are given in Table 1, together with the antidot hole diameter and the corresponding aspect ratio.

Table 1. Effective INP anisotropy constants, K_{eff}, for $Dy_{13}Fe_{87}$ thin films and antidot arrays with different thicknesses (t), antidot hole diameters (d) and antidot aspect ratios (r).

t (nm)	d (nm)	r	K_{eff} (erg/cm^3)	
			Antidots	Thin Film
20	56	3.3	2.7×10^5	3.5×10^6
30	45	2.2	-1.2×10^6	7.4×10^6
50	15	0.9	1.2×10^6	8.0×10^6

For the $t = 20$ nm sample, the aspect ratio, r, takes values of around three, indicating that the antidots display larger diameter than thickness, and therefore, the magnetization preferred direction remains in plane. Nevertheless, the effect of antidots in this sample becomes evident, if we compare the K_{eff} values for 20 nm thick antidot and thin film samples shown in Table 1. The difference in one order of magnitude between them can be ascribed to the competition between INP and OOP anisotropies. In contrast, for the 30 nm thick sample, the aspect ratio is closer to two, that is, the thickness of the thin film is near to the average antidot diameters. Therefore, the magnetostatic energy accumulated by the surface poles, in the film plane, and that due to magnetic poles on the nanohole surfaces starts to be of the same order of magnitude. Thus, the resulting magnetization lies in the OOP direction,

Nanomaterials **2018**, *8*, 227

as the thickness, t, increases, in order to reduce the whole magnetostatic energy of the system [32,33], leading to a negative value of K_{eff}. Finally, when the film thickness is further increased, the antidot hole diameter is greatly reduced, due to the deposition of material in the pore walls. As a consequence, the antidot film behaves similarly to the continuous thin film, as evidenced by the positive sign of K_{eff}, which takes values of the same order of magnitude than for the continuous thin films.

4. Conclusions

In this work, we report on the transfer of the easy magnetization axis from the in plane, in the case of $Dy_{13}Fe_{87}$ thin film, to out of plane, for the antidot arrays film with the same thickness condition.

The antidot arrays introduce a drastic change in the morphology and magnetic behavior of the magnetic thin film if we compare it with that one of the continuous thin film. It increases the magnetostatic energy associated to the antidot arrays, if the magnetization lays in the film plane. This energy might be due to the appearance of magnetic poles on the antidot surfaces in competition with the magnetostatic energy due to the magnetic poles on the film surface when the magnetization is OOP. Obviously, the magnetostatic energy associated to the antidot array, increases with the film thickness. When the layer thickness increases well enough to counterbalance the energy associated to the magnetic poles on the film surface, the preferred direction of magnetization should change from INP to OOP direction. This effect has been observed for $t = 30$ nm, whereas for samples with lower or higher film thickness, the magnetization remains INP, which is ascribed to the high value of antidots aspect ratio and to the reduction in hole size with the increase in layer thickness, respectively.

Additionally, by suitably controlling the antidot aspect ratio parameter, it also allows for tailoring the magnetization process of the magnetic materials and controlling the direction of easy magnetization axis from the in plane or out of plane directions.

Furthermore, the change of INP to an OOP magnetization signal may open up new directions in magnetic sensing or spintronic applications at the nanoscale by combining two devices that need different orientations of the magnetic signal. Finally, the transformation of in-plane magnetic information to an out-of-plane magnetic signal may advance 2D magnetic logic to the third dimension.

Acknowledgments: This work was supported by the Ministerio de Economia y Competitividad (MINECO) from Spain, under grants MAT2013-48054-C2-2-R, MAT2016-76824-C3-1-R and MAT2016-76824-C3-3-R; together with the Principado de Asturias (Spain) under grant FC-15-GRUPIN14-085. M. Jaafar wishes also to acknowledge her funding grant from Spanish MINECO under MAT2015-73775-JIN research project. The scientific support from the SCTs of the University of Oviedo is also gratefully recognized.

Author Contributions: A.F. and V.M.P. conceived and designed the experiments; V.V. synthesized the nanoporous alumina membranes and performed SEM characterization; M.S. and A.F. deposited the antidot thin films; M.S. and V.V. performed the magnetic characterization; A.I., M.J. and A.A. carried out the AFM/MFM studies; M.S., V.M.P. and A.F. analyzed the data. All authors have contributed in co-writing of the manuscript and gave approval to the final revision.

Conflicts of Interest: The authors declare no conflict of interest. The founding sponsors had no role in the design of the study; in the collection, analyses, or interpretation of data; in the writing of the manuscript, and in the decision to publish the results.

References

1. Stamps, R.L.; Breitkreutz, S.; Akerman, J.; Chumak, A.V.; Otani, Y.; Bauer, G.E.W.; Thiele, J.-U.; Bowen, M.; Majetich, S.A.; Klaui, M.; et al. The 2014 Magnetism Roadmap. *J. Phys. D Appl. Phys.* **2014**, *47*, 333001. [CrossRef]

2. Zhang, S.; Gilbert, I.; Nisoli, C.; Chern, G.W.; Erickson, M.J.; O'Brien, L.; Leighton, C.; Lammenrt, P.E.; Crespi, V.H.; Schiffer, P. Crystallites of magnetic charges in artificial spin ice. *Nature* **2013**, *500*, 553–557. [CrossRef] [PubMed]

3. Heyderman, L.J. Artificial spin ice: Crystal-clear order. *Nat. Nanotechnol.* **2013**, *8*, 705–706. [CrossRef] [PubMed]

4. Laguna, M.F.; Balseiro, C.A.; Domínguez, D.; Nori, F. Vortex structure and dynamics in kagomé and triangular pinning potentials. *Phys. Rev. B* **2001**, *64*, 104505. [CrossRef]

5. Abad, A.; Corma, A.; García, H. Supported gold nanoparticles for aerobic, solventless oxidation of allylic alcohols. *Pure Appl. Chem.* **2007**, *79*, 1847–1854. [CrossRef]

6. Mengotti, E.; Heyderman, L.J.; Rodriguez, A.F.; Nolting, F.; Hugli, R.V.; Braun, H.B. Real-space observation of emergent magnetic monopoles and associated Dirac strings in artificial kagome spin ice. *Nat. Phys.* **2011**, *7*, 68–74. [CrossRef]

7. Ünal, A.A.; Valencia, S.; Radu, F.; Marchenko, D.; Merazzo, K.J.; Vázquez, M.; Sánchez-Barriga, J. Ferrimagnetic DyCo5 Nanostructures for Bits in Heat-Assisted Magnetic Recording. *Phys. Rev. Appl.* **2016**, *5*, 064007. [CrossRef]

8. Cowburn, R.P.; Adeyeye, A.O.; Bland, J.A.C. Magnetic domain formation in lithographically defined antidot Permalloy arrays. *Appl. Phys. Lett.* **1997**, *70*, 2309–2311. [CrossRef]

9. Metaxas, P.J.; Sushruth, M.; Begley, R.A.; Ding, J.; Woodward, R.C.; Maksymov, I.S.; Albert, M.; Wang, W.; Fangohr, H.; Adeyeye, A.O.; et al. Sensing magnetic nanoparticles using nano-confined ferromagnetic resonances in a magnonic crystal. *Appl. Phys. Lett.* **2015**, *106*, 232406. [CrossRef]

10. Yu, H.; Duerr, G.; Huber, R.; Bahr, M.; Schwarze, T.; Brandl, F.; Grundler, D. Omni directional spin-wave nanograting coupler. *Nat. Commun.* **2013**, *4*, 2702. [CrossRef] [PubMed]

11. Zhang, W.; Li, J.; Ding, X.; Pernod, P.; Tiercelin, N.; Song, Y. Tunable Magneto-Optical Kerr Effect of Nanoporous Thin Films. *Sci. Rep.* **2017**, *7*, 2888. [CrossRef] [PubMed]

12. Coïsson, M.; Federica Celegato, F.; Gabriele Barrera, G.; Gianluca Conta, G.; Alessandro Magni, A.; Paola Tiberto, P. Bi-Component Nanostructured Arrays of Co Dots Embedded in Ni80Fe20 Antidot Matrix: Synthesis by Self-Assembling of Polystyrene Nanospheres and Magnetic Properties. *Nanomaterials* **2017**, *7*, 232. [CrossRef] [PubMed]

13. Haering, F.; Wiedwald, U.; Nothelfer, S.; Koslowski, B.; Ziemann, P.; Lechner, L.; Wallucks, A.; Lebecki, K.; Nowak, U.; Gräfe, J.; et al. Switching modes in easy and hard axis magnetic reversal in a self-assembled antidot array. *Nanotechnology* **2013**, *24*, 465709. [CrossRef] [PubMed]

14. Prida, V.M.; Salaheldeen, M.; Pfitzer, G.; Hildalgo, A.; Vega, V.; González, S.; Teixeira, J.M.; Fernández, A.; Hernando, B. Template Assisted Deposition of Ferromagnetic Nanostructures: From Antidot Thin Films to Multisegmented Nanowires. *Acta Phys. Pol. A* **2017**, *131*, 822–827. [CrossRef]

15. Gräfe, J.; Schütz, G.; Goering, E.J. Coercivity scaling in antidot lattices in Fe, Ni, and NiFe thin films. *J. Magn. Magn. Mater.* **2016**, *419*, 517–520. [CrossRef]

16. Masuda, H.; Fukuda, K. Ordered metal nanohole arrays made by a two-step replication of honeycomb structures of anodic alumina. *Science* **1995**, *268*, 1466–1468. [CrossRef] [PubMed]

17. Prida, V.M.; Pirota, K.R.; Navas, D.; Asenjo, A.; Hernández-Vélez, M.; Vázquez, M. Self-organized magnetic nanowire arrays based on alumina and titania templates. *J. Nanosci. Nanotechnol.* **2007**, *7*, 272–285. [CrossRef] [PubMed]

18. Béron, F.; Pirota, K.R.; Vega, V.; Prida, V.M.; Fernández, A.; Hernando, B.; Knobel, M. An effective method to probe local magnetostatic properties in a nanometric FePd antidot array. *New J. Phys.* **2011**, *13*, 013035. [CrossRef]

19. López-Antón, R.; Vega, V.; Prida, V.M.; Fernández, A.; Pirota, K.R.; Vázquez, M. Magnetic properties of hexagonally ordered arrays of Fe antidots by vacuum thermal evaporation on nanoporous alumina templates. *Solid State Phenom.* **2009**, *152–153*, 273–276. [CrossRef]

20. Rahman, M.T.; Shams, N.N.; Wang, D.S.; Lai, C.H. Enhanced exchange bias in sub-50-nm IrMn/CoFe nanostructure. *Appl. Phys. Lett.* **2009**, *94*, 082503. [CrossRef]

21. Xiao, Z.L.; Han, C.Y.; Welp, U.; Wang, H.H.; Vlasko-Vlasov, V.K.; Kwok, W.K.; Miller, D.J.; Hiller, J.M.; Cook, R.E.; Willing, G.A.; et al. Nickel antidot arrays on anodic alumina substrates. *Appl. Phys. Lett.* **2002**, *81*, 2869. [CrossRef]

22. Mansuripur, M.J. Analysis of multilayer thin-film structures containing magneto-optic and anisotropic media at oblique incidence using 2 × 2 matrices. *Appl. Phys.* **1990**, *67*, 6466. [CrossRef]

23. Gräfe, J.; Haering, F.; Tietze, T.; Audehm, P.; Weigand, M.; Wiedwald, U.; Ziemann, P.; Gawroński, P.; Schütz, G.; Goering, E.J. Perpendicular magnetisation from in-plane fields in nano-scaled antidot lattices. *Nanotechnology* **2015**, *26*, 225203. [CrossRef] [PubMed]

24. Cowburn, R.P.; Adeyeye, A.O.; Bland, J.A.C. Magnetic switching and uniaxial anisotropy in lithographically defined anti-dot Permalloy arrays. *J. Magn. Magn. Mater.* **1997**, *173*, 193–201. [CrossRef]

25. Ambrose, M.C.; Stamps, R.L. Magnetic stripe domain pinning and reduction of in-plane magnet order due to periodic defects in thin magnetic films. *J. Magn. Magn. Mater.* **2013**, *344*, 140–147. [CrossRef]

26. Van de Wiele, B.; Manzin, A.; Vansteenkiste, A.; Bottauscio, O.; Dupre, L.; De Zutter, D. A micromagnetic study of the reversal mechanism in permalloy antidot arrays. *J. Appl. Phys.* **2012**, *111*, 053915. [CrossRef]

27. Weller, D.; Stohr, J.; Nakajima, R.; Carl, A.; Samant, M.G.; Chappert, C.; Megy, R.; Beauvillain, P.; Veillet, P.; Held, G.A. Microscopic origin of magnetic anisotropy in Au/Co/Au probed with X-ray magnetic circular dichroism. *Phys. Rev. Lett.* **1995**, *75*, 3752–3755. [CrossRef] [PubMed]

28. Baudelet, F.; Lin, M.T.; Kuch, W.; Meinel, K.; Choi, B.; Schneider, C.M.; Kirschner, J. Perpendicular anisotropy and spin reorientation in epitaxial Fe/Cu$_3$Au(100) thin films. *Phys. Rev. B* **1995**, *51*, 12563–12578. [CrossRef]

29. Navas, D.; Hernández-Vélez, M.; Vázquez, M.; Lee, W.; Nielsch, K. Ordered Ni nanohole arrays with engineered geometrical aspects and magnetic anisotropy. *Appl. Phys. Lett.* **2007**, *90*, 192501. [CrossRef]

30. Vavassori, P.; Gubbiotti, G.; Zangari, G.; Yu, C.T.; Yin, H.; Jiang, H.; Mankey, G.J. Lattice symmetry and magnetization reversal in micron-size antidot arrays in Permalloy film. *J. Appl. Phys.* **2002**, *91*, 7992. [CrossRef]

31. Wang, C.C.; Adeyeye, A.O.; Singh, N. Magnetic antidot nanostructures: Effect of lattice geometry. *Appl. Phys. Lett.* **2006**, *88*, 222506. [CrossRef]

32. Merazzo, K.J.; Castan-Guerrero, C.; Herrero-Albillos, J.; Kronast, F.; Bartolome, F.; Bartolome, J.; Sese, J.; del Real, R.P.; Garcia, L.M.; Vazquez, M. X-ray photoemission electron microscopy studies of local magnetization in Py antidot array thin films. *Phys. Rev. B* **2012**, *85*, 184427. [CrossRef]

33. Merazzo, K.J.; Leitao, D.C.; Jimenéz, E.; Araujo, J.P.; Camarero, J.; del Real, R.P.; Asenjo, A.; Vázquez, M. Geometry-dependent magnetization reversal mechanism in ordered Py antidot arrays. *J. Phys. D Appl. Phys.* **2011**, *44*, 505001. [CrossRef]

nanomaterials

MDPI

Article

Crosslinked Polymer Ionic Liquid/Ionic Liquid Blends Prepared by Photopolymerization as Solid-State Electrolytes in Supercapacitors

Po-Hsin Wang [1], Tzong-Liu Wang [1], Wen-Churng Lin [2], Hung-Yin Lin [1], Mei-Hwa Lee [3] and Chien-Hsin Yang [1,*]

[1] Department of Chemical and Materials Engineering, National University of Kaohsiung, Kaohsiung 81148, Taiwan; p5802341@gmail.com (P.-H.W.); tlwang@nuk.edu.tw (T.-L.W.); linhy@nuk.edu.tw (H.-Y.L.)
[2] Department of Environmental Engineering, Kun Shan University, Tainan 71070, Taiwan; linwc@mail.ksu.edu.tw
[3] Department of Materials Science and Engineering, I-Shou University, Kaohsiung 84001, Taiwan; mhlee@isu.edu.tw
* Correspondence: yangch@nuk.edu.tw; Tel.: +886-7-5919420

Received: 8 February 2018; Accepted: 4 April 2018; Published: 7 April 2018

Abstract: A photopolymerization method is used to prepare a mixture of polymer ionic liquid (PIL) and ionic liquid (IL). This mixture is used as a solid-state electrolyte in carbon nanoparticle (CNP)-based symmetric supercapacitors. The solid electrolyte is a binary mixture of a PIL and its corresponding IL. The PIL matrix is a cross-linked polyelectrolyte with an imidazole salt cation coupled with two anions of Br^- in PIL-M-(Br) and $TFSI^-$ in PIL-M-(TFSI), respectively. The corresponding ionic liquids have imidazolium salt cation coupled with two anions of Br^- and $TFSI^-$, respectively. This study investigates the electrochemical characteristics of PILs and their corresponding IL mixtures used as a solid electrolyte in supercapacitors. Results show that a specific capacitance, maximum power density and energy density of 87 and 58 $F \cdot g^{-1}$, 40 and 48 $kW \cdot kg^{-1}$, and 107 and 59.9 $Wh \cdot kg^{-1}$ were achieved in supercapacitors based on (PIL-M-(Br)) and (PIL-M-(TFSI)) solid electrolytes, respectively.

Keywords: supercapacitor; polymer ionic liquid; ionic liquid; solid electrolyte; photopolymerization

1. Introduction

Green power generation and energy storage have been the most attractive subject of research in recent years. Supercapacitors are an energy storage device. Supercapacitors are often used in high power output or energy storage systems such as starters, hybrid cars and uninterruptible power supplies [1–3].

Supercapacitors have electric double layer capacitors (EDLC) and pseudocapacitors. The former uses a physical method (Coulombs static force) to store charges, whereas the latter uses a chemical method (redox reaction) to store the charges. Asymmetric devices are generally composed of the pseudocapacitor anode [4] and the electric double layer cathode. The main materials are carbon materials, metal oxides, and conductive polymers. Carbon is the most common electrode material for supercapacitors due to low cost, high specific surface area, good conductivity, and high chemical stability. Carbon materials are mainly used in EDLC, but functional groups are involved on their surface that would lead to the appearance of pseudo-capacitance characteristics [5].

In supercapacitors, the electrolyte needs to provide a sufficient concentration of positive and negative ions, good ionic conductivity, and no chemical reaction with the packaging material. In addition, the electrolyte also determines the operation potential window of a supercapacitor. In carbon-based supercapacitors using 1 M sulfuric acid and 6 M potassium hydroxide [6] at a charge–discharge current density of 0.04 A·g^{-1}, the potential windows only reached 0.6 and 1 V, and the specific capacitances of 242 and 208 F·g^{-1}, respectively. The potential operated above 1.23 V will cause water to be decomposed, insinuating the potential-window limitation of 1 V in the aqueous electrolyte. Moreover, aqueous electrolytes often contain a strong acid (H$_2$SO$_4$) and base (KOH), leading to electrode corrosion. Organic electrolyte has a higher operating potential (about 2 V) in relation to the aqueous system, so that this system can store more charge. In addition, organic electrolyte not only has lower corrosion than the aqueous electrolyte, but also less reactivity with the electrode. However, organic electrolyte has a relatively high resistance resulting from poor ion conductivity. Ionic liquids (ILs) have a wide range of operating potential window that allow them to store more charges with a higher energy density, chemical stability at high temperature, and environmental friendliness. Solid-state electrolytes are mostly prepared by incorporating polymers in a liquid electrolyte. The ionic conductivity of the solid electrolyte is lower than that of the above two systems, mainly because the ion transport in the solid electrolyte is hard, causing a large internal resistance, and the separator can be omitted in the device.

Ionic liquid can be prepared with the desired characteristics according to different needs, because it has a great variety of cation and anion combinations [7]. Izmaylova et al. [8] used 1-methyl-3-butylimidazolium tetrafluoroborate ([MBIM] [BF$_4$]) as the electrolyte in a carbon-based symmetric supercapacitor, obtaining a specific energy density and specific power density of 4.1 Wh·kg^{-1} and 1.7 Wk·g^{-1}, respectively. A symmetric supercapacitor was made using activated carbon fiber cloth as the electrodes and mixed 1-ethyl-3-methylimidazolium bromide and 1-ethyl-3-methylimidazolium tetrafluoroborate as the electrolyte ([EMIm] [Br]/[EMIm] [BF$_4$]). The potential window was up to 2 V, and the specific capacitances were 59 F·g^{-1} at a charge–discharge current density of 0.1 A·g^{-1}. The performance of electrolyte-containing bromide ion is better than the electrolyte without bromide ion [9]. However, there is a drawback—electrolyte leakage—which makes IL extremely troublesome to package the supercapacitors. To solve this problem, a solid or quansi-solid state electrolyte is substituted for IL. In 2014, a polymer electrolyte of tetramethoxysilane and formic acid was prepared employing the sol–gel reaction and then mixed with 1-ethyl-3-methylimidazolium bis(trifluoromethanesulfonimide) ([EMIm] [TFSI]), which has been utilized in a carbon-based symmetric supercapacitor with a potential window of 3 V at a charge–discharge density of 1 A·g^{-1} [10] and the specific capacitance of 48 F·g^{-1} with 100% capacitance retention after 10,000 cycles. Poly (ionic liquids) (PILs) are made by the polymerization of corresponding ionic liquid monomers. Most PILs are a polycation in which the cationic groups bear on the polymer main chain. Except for common pyridinium, pyrrolidonium, and imidazolium, there still are vinyl, styrenic, and methacrylic groups bearing on the monomers through free radical polymerization to obtain the corresponding polymers. Recently, cross-linked poly-4-vinylphenol (c-P4VPh) was employed as a polymer matrix and mixed 1-ethyl-3-methylimidazolium bis(trifluoromethylsulfonyl) imide ([EMI] [TFSI]) as a polymer electrolyte, which was involved in a porous carbon-based symmetric supercapacitor [11] with a potential window of 4 V at a charge–discharge density of 1 mA·cm^{-2} and a specific capacitance of 172.45 F·g^{-1}, an energy density of 72.26 Wh·kg^{-1}, and a power density of 1696.56 W·kg^{-1}. More recently, photopolymerization was employed to polymerize vinylimidazole monomers containing bromine anions in PILs [12–14], obtaining non-crystalline polymers which are very suitable for use as solid electrolytes [14].

In this work, we hope to prepare a solid electrolyte instead of the liquid electrolyte and separator in a carbon nanoparticle (CNP)-based supercapacitor. We chose carbon nanoparticle (CNP) as the electrode material of the supercapacitor due to low cost and stability. To enhance the ionic conductivity of solid electrolytes, a binary component of IL soaked into PIL was employed as a solid state electrolyte. Here, butane-substituted vinylimidazolium salt, 1-methyl-3-butylimidazolium as the cation and bromide as the anion, was photopolymerized to form a PIL. This has the advantages of good thermal stability, high electrochemical stability and high ionic conductivity to achieve a high performance carbon nanoparticle (CNP)-based supercapacitor. To increase the potential window enhancing the energy density, the bromide anion was ion-exchanged by bis(trifluoromethane sulfonamide anion

(TFSI^{-1}) in ionic liquid, whereas a cross-linked PIL was prepared as the solid electrolyte with high mechanical strength, retaining more IL.

2. Materials and Methods

2.1. Chemicals

1-Bromobutane (Alfa Aesar, Tewksbury, MA, USA), 1,4-dibromobutane (Alfa Aesar, Tewksbury, MA, USA), 1-vinylimidazole (Alfa Aesar, Tewksbury, MA, USA), 1-methylimidazole (Alfa Aesar, Tewksbury, MA, USA), lithium bis(trifluoromethane sulfonamide) (LiTFSI, Acros, Geel, Belgium), photoinitiators: diphenyl (2,4,6-trimethylbenzoyl)-phosphine oxide (TPO, Aldrich, Milwaukee, WI, USA), 1-hydroxy-cyclohexyl-phenyl-ketone (Irgacure 184, Ciba, Tarrytown, NY, USA), 2-methyl-1-[4-(methylthio)phenyl]-2-(4-morpholinyl)-1-propanone (Irgacure 907, Ciba, Tarrytown, NY, USA), poly(vinylidene fluoride) (PVDF, Solf®PVDF 6020, Solvay, Bruxelles, Belgium), 1-methyl-2-pyrrolidone (NMP, Riedel-de Haën, Seelze, Germany), carbon nanoparticles (CNP, ~15 nm, UniRegion Bio-Tech, Taoyuan, Taiwan), and activated carbon (AC, ACS-2930, China steel chem. Co., Kaohsiung, Taiwan) were used as received.

2.2. Preparation of Ionic Liquid

The chemical reactions involved in the preparation of the materials are sketched in Table 1.

Table 1. Preparation process of the ionic liquid monomer.

Component	Reaction Scheme
MBIB	
BVIB	
BDVIB	
MBIT	

Table 1. *Cont.*

Component	Reaction Scheme
BVIT	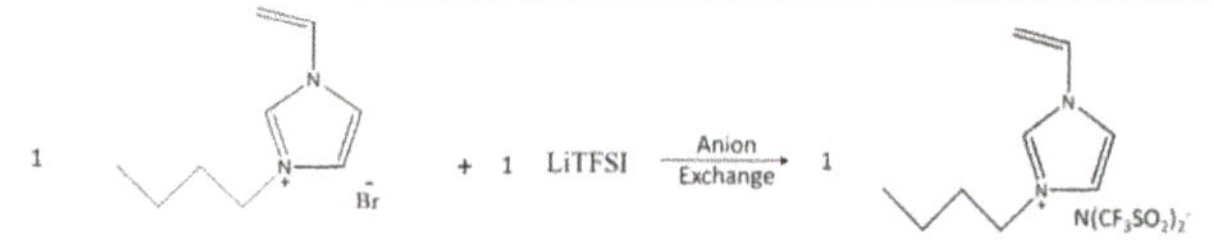
BDVIT	
PIL-(Br)	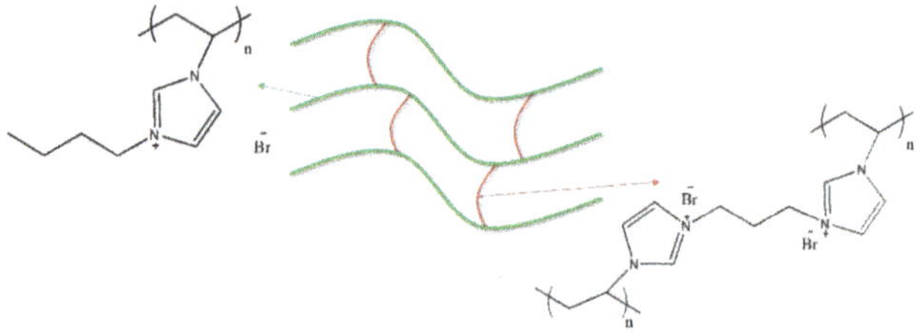
PIL-(TFSI)	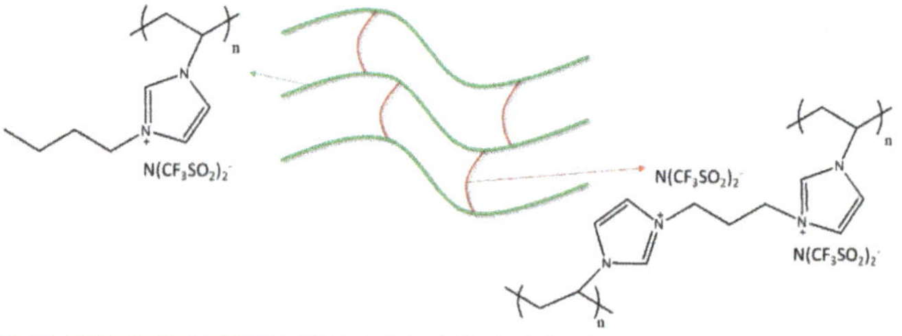

2.2.1. Preparation of Ionic Liquid Monomer

The general synthetic procedures for ionic liquid monomers (ILMs) followed the literature [15]. To prepare 1-methyl-3-butylimidazolium bromide (MBIB), 0.1 mole of 1-methylimidazole, 0.1 mole of 1-bromobutane and 30 mL of methanol were loaded into a 100 mL reactor. The mixture was stirred at 60 °C for 15 h. After cooling down, the reaction mixture was added dropwise into 1 L of diethyl ether to extract the product. A yellow liquid product (1-methyl-3-butylimidazolium bromide, MBIB) was collected and completely dried at 60 °C under vacuum.

To prepare 3-*n*-butyl-1-vinylimidazolium bromide (BVIB), 0.1 mole of 1-vinylimidazole was used to replace 1-methylimidazole. The other chemicals were the same as for the preparation of MBIB, and the preparation procedure also followed the preparation of MBIB.

To prepare 1,4-butanediyl-3,3′-bis-l-vinylimidazolium dibromide (BDVIB), 0.2 mole of 1-vinylimidazole and 0.1 mole of 1,4-dibromobutane were used to replace 0.1 mole of 1-methylimidazole and 0.1 mole of 1-bromobutane, respectively. Other chemicals were the same as the preparation of MBIB, and the preparation procedure also followed the preparation of MBIB.

To prepare MBIT, 0.01 mole of MBIB, 0.01 mole of LiTFSI, and 8 mL of distilled water were mixed and stirred in a three-necked flask at 80 °C for 24 h under nitrogen purge. Then, 15 mL of distilled water solution was added to the above reaction solution and kept stirring to cool at room temperature. The solution was extracted using CH$_2$Cl$_2$ to extract, and then repeatedly washed with distilled water. A yellow liquid product was obtained after being set in a vacuum oven for 24 h.

The preparation procedures of BVIT were the same as that of MBIT rather than MBIB by VBIB. The preparation procedures of BDVIT were the same as that of MBIT rather than MBIB by BDVIB and using 0.02 mole of LiTFSI.

2.2.2. Preparation of PIL

To prepare crosslinked PIL-(Br), 16.77 mmole of BVIB, 2.93 mmole of BDVIB, 150 mg of mixed photoinitiators (60 mg TPO + 45 mg Irgacure 184 + 45 mg Irgacure 907), and 1.5 g of acetone were mixed at 50 °C until completely dissolved. Acetone in the mixture was removed by purging with dry argon at 60 °C for 24 h. After cooling down to room temperature, the reaction mixture was cast on a 50 cm^2 area polyimide plate. This thin IL film was exposed to ultraviolet light irradiation of 125 W (emission peak at 365 nm, Ausbond) keeping a distance of 10 cm from the film to the mercury bulb for 10 min, the power level was 8 mW/cm^2 measured at 365 nm, performing photopolymerization to generate a film product on the polyimide substrate. The hypothesized molecular structures of PIL-(Br) and PIL-(TFSI) are proposed in Table 1.

To prepare the PIL/IL blend (PIL-M-(Br)), 4.3 mmole of MBIB, 16.77 mmole of BVIB, 2.93 mmole of BDVIB, 150 mg of mixed photoinitiators (60 mg TPO + 45 mg Irgacure 184 + 45 mg Irgacure 907), and 1.5 g of acetone were mixed at 50 °C until completely dissolved. Acetone in the mixture was removed by purging with dry argon at 60 °C for 24 h. After cooling down to room temperature, the reaction mixture was cast on a 50 cm^2 area polyimide plate. This thin IL film was exposed to ultraviolet light irradiation of 125 W (emission peak at 365 nm, Ausbond) keeping a distance of 10 cm from the film to the mercury bulb for 10 min, the power level was 8 mW/cm^2 measured at 365 nm, performing photopolymerization to generate a film product on the polyimide substrate.

The preparation procedures of crosslinked PIL-(TFSI) were the same as that of crosslinked PIL-(Br) rather than BDVIB by BDVIT and VBIB by VBIT, respectively. The preparation procedures of PIL/IL blend (PIL-M-(TFSI)) were the same as that of PIL/IL blend (PIL-M-(Br)) rather than BDVIB by BDVIT, VBIB by VBIT, and MBIB by MBIT, respectively.

2.3. Electrode Preparation

Graphite paper (20 × 10 mm) was ground to a smooth surface using coarse and fine sandpaper in a sequence, washed using 0.5 M sulfuric acid to clean the surface, and then transferred to dry in the oven at 120 °C for 2 h obtaining a graphite-paper current collector. Activated carbon, carbon nanoparticles, and PVDF were mixed and ground in an agate bowl with a weight ratio of 7:1:1, and then an appropriate amount of NMP was added and pulverized to a slurry paste. An appropriate amount of the paste was dropped on the polish graphite paper (10 × 10 mm) and spin-coated to form a uniform film. Then the above film was baked in a circulation oven at 70 °C for 24 h to obtain the carbon electrode. The symmetric capacitors were sandwiched by two working electrodes that were separated by polymer ionic liquid/ionic liquid (PIL-M-(Br) or PIL-M-(TFSI)) blend films as the electrolyte and separator with a thickness of ca. 0.25 mm. IL electrolytes of MBIB and MBIT were injected into a bipolar CNP-electrode cell with a porous polypropylene separator (~0.25 mm thickness).

2.4. Characterization and Measurements

Infrared spectra were recorded on a Fourier transform infrared spectroscopy (FTIR) spectrometer (Agilent Technologies, Cary 630, Santa Clara, CA, USA) to check the functional groups on polymer films using an attenuated total reflection (ATR) cell. Film surface morphology was observed on field emission scanning electron microscopy (FESEM, Hitachi, S4800, Tokyo, Japan). The PIL film samples were put on aluminum film prior to imaging. ^{1}H nuclear magnetic resonance (^{1}H NMR) measurements were carried out at room temperature using a Bruker AV400 FT-NMR spectrometer (Billerica, MA, USA) operating at 400 MHz. Chloroform-d$_1$ or dimethylsulfoxide (DMSO)-d$_6$ were used as solvents. Differential scanning calorimetry (DSC) measurements were performed on a TA instrument (SDT-Q600, New Castle, DE, USA). The samples were first heated up to 100 °C, and then the samples were cooled from +100 to −50 °C at a cooling rate of 10 °C/min in a cooling process, where they were kept at −50 °C for 2 min. Finally, the samples were reheated to 295 °C at a heating rate of 10 °C/min. The glass transition point of PILs was determined by the heating curve. Thermal gravitational analysis (TGA) was performed on a TA instrument (SDT-Q100, New Castle, DE, USA).

2.5. Electrochemical Characterization

All of the electrochemical experiments were investigated using an AUTOLAB PGSTAT302N electrochemical work station (Metrohm Autolab, Utrecht, The Netherlands). A typical three-electrode cell, filled with electrolyte and equipped with a working electrode, platinum foil as a counter electrode and an Ag/AgCl reference electrode, was employed to measure the electrochemical properties of the AC-CNP carbon-coated working electrode and its symmetric device at 25 °C. The cyclic voltammograms (CVs) of the AC-CNP carbon-coated electrode were recorded by varying the scan rates: 5, 10 mV·s^{-1} over the potential range of −0.2–0.8 V for the three-electrode cell and 10, 25, 50, 75, 100 and 125 mV·s^{-1} over the potential range of 0–0.8 V for the symmetric supercapacitors. Measurements of electrochemical impedance spectroscopy (EIS) were carried out on the bipolar CNP-electrode cell using an alternating current (AC) bias of 0.1 V with a signal of 10 mV over the frequency range of 0.1 Hz to 100 kHz at 25 °C. Solid-electrolyte films (~0.25 mm thickness) of PIL-M-(Br) and PIL-M-(TFSI) were directly sandwiched between the bipolar CNP-electrodes. IL electrolytes of MBIB and MBIT were injected into a bipolar CNP-electrode cell with a porous polypropylene separator (~0.25 mm thickness). The galvanostatic charge–discharge curves were obtained over the potential range of 0–3.0 V for various current densities (0.5, 1, 1.5, 2, 2.5, 5 and 10 A·g^{-1}) in symmetric devices. A current density of 1, 5 A·g^{-1} was used in the symmetric cell.

3. Results and Discussion

3.1. Morphology of PILs Derived from Photopolymerization

Figure 1a shows SEM image of the PIL-(Br) film, exhibiting an uneven surface with fine cracks and bumps. These features could be due to shrinking upon polymerization and thus suggest the presence of considerable attractive intermolecular interactions between polymeric chains in the forming films. The SEM image of the PIL-M-(Br) film exhibited a smoother surface with a few fine holes in comparison with the compared PIL-(Br) film as shown in Figure 1b. This result suggests that the fine drops of MBIB liquid phase are present in the generating polymer film during photopolymerization in the binary mixture of BVIB (Br$^-$), BDVIB (Br$^-$), and MBIB (Br$^-$). On the other hand, a cross-linked network structure of PIL-(TFSI) polymer can be observed with a uniform distribution of tangled lines in Figure 1c. Figure 1d shows the image of PIL-M-(TFSI), exhibiting a smooth surface with many holes of different sizes. This result implies that the MBIB liquid drop phase is also present in the generating polymer film during photopolymerization in the binary mixture of BDVIB (TFSI$^-$), BDVIB (TFSI$^-$), and MBIB (TFSI$^-$).

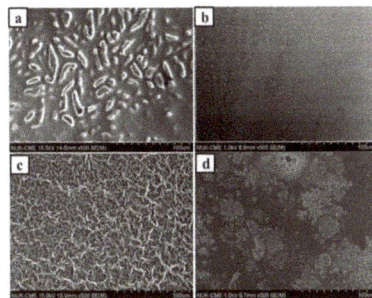

Figure 1. Scanning electron microscopy (SEM) images of (**a**) polymer ionic liquid (PIL)-(Br); (**b**) PIL-M-(Br); (**c**) PIL-(TFSI); and (**d**) PIL-M- (TFSI) films.

3.2. FTIR Spectra of PILs Derived from Photopolymerization

Figure 2 shows the FTIR spectra of PIL-(Br), PIL-M-(Br), PIL-(TFSI), and PIL-M-(TFSI) and Table 2 lists the assignment of functional groups above the polyelectrolytes. The stretching of C–H (SP^2) presents at 3077 cm^{-1}, which is contributed by the C=C–H bond on the imidazole ring. The stretching of C–H (SP^3) presents at the three peaks of 2981, 2932 and 2869 cm^{-1}, corresponding to the C–C–H bond on the imidazole ring. The stretching of C=C and C=N on the aromatic ring presents at 1650 and 1546 cm^{-1}, corresponding to the C=C and C=N bonds, respectively, on the imidazole ring. The bending of the C–H (SP^3) at 1457 cm^{-1} corresponds to CH$_3$ substituent at the end of the imidazole ring. The C–N bending on the aromatic imidazole ring presents at 1151 cm^{-1}. The peaks in the region of 600–900 cm^{-1} result from the stretching of alkyl halogen bonding or the presence of halide ions. These results are consistent with the assignment of the molecular structure of PIL-(Br) and PIL-M-(Br) [15]. Also note that there is a peak at 3141 cm^{-1} corresponding to the bending of the C–H bond on the imidazole ring. The peaks at 2981 and 2854 cm^{-1} represent the stretching of symmetric and asymmetric C–H (SP^3) at the end substituent on the imidazole ring. The stretching at 1654 cm^{-1} corresponds to the C=N bond on the imidazole ring. The bending of the NH bond on the imidazole ring presents at 1552 cm^{-1}. The bending of the C–H (SP^3) at 1460 cm^{-1} corresponds to the CH$_3$ at the end substituent on the imidazole ring. The peaks at 1345 and 1126 cm^{-1} represent symmetric and asymmetric SO$_2$ bond stretching on the TFSI anion. The C–N stretching on the aromatic imidazole ring presents at 1187 cm^{-1} on the imidazole ring. The bending of symmetric CF$_3$ bonding presents at 1046 cm^{-1} in the TFSI anion; the peak of 849 cm^{-1} corresponds to the stretching of an asymmetric S–N–S bond in the TFSI anion.

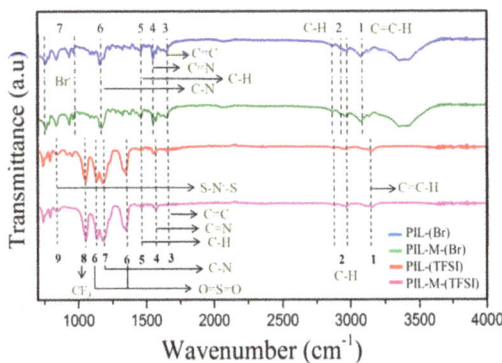

Figure 2. Fourier transform infrared spectroscopy (FTIR) spectra of PIL-(Br), PIL-M-(Br), PIL-(TFSI), and PIL-M-(TFSI).

Table 2. IR-ATR observed wavelengths, in cm^{-1}, of the absorption bands detected in the spectra of PIL-(Br), PIL-M-(Br), PIL-(TFSI) and PIL-M-(TFSI).

No.	PIL-(Br)/PIL-M-(Br)	Assignment	PIL-(TFSI)/PIL-M-(TFSI)	Assignment
1	3077	C–H bending	3141	C–H bending
2	2981 2932 2869	C–H streching	2981 2854	C–H streching
3	1650	C=C strehing	1654	C=N streching
4	1546	C=N bending	1552	C=N bending
5	1457	C–H bending	1460	C–H bending
6	1151	C–N bending	1345 1126	O=S=O streching
7	600–900	bromide	1187	C–N streching
8	-	-	1046	CF bending
9			849	S–N–S

3.3. Thermal Gravity Analysis of PILs Derived from Photopolymerization

Figure 3a shows differential scanning calorimetric analysis of PIL-(TFSI) and PIL-(Br), revealing that T_g (15.5 °C) of PIL-(TFSI) is lower than that (31 °C) of PIL-(Br). Also note that T_m (126 °C) of PIL-(TFSI) is lower than that (135 °C) of PIL-(Br). These features suggest that the molecular packing in the solid state is more efficient in presence of a Br$^-$ anion than in the case of TFSI$^-$. These two anions are quite different in terms of steric hindrance, symmetry and charge density, properties typically involved in solid-state lattice enthalpy and entropy. Figure 3b shows the thermal gravity analysis (TGA) of PIL-(Br), PIL-M-(Br), PIL-(TFSI), and PIL-M-(TFSI). The weight loss of the PIL-(Br) and PIL-M-(Br) samples was about 17% at 200 °C due to the evaporation of water, whereas the major weight losses occurred at 305 and 289 °C, respectively, mainly due to the thermal decomposition of PIL-(Br) and PIL-M-(Br). It can be seen that the presence of soaked MBIB resulted in a lower thermal stability of the material. Note that the major thermal decomposition temperature of PIL-(TFSI) and PIL-M-(TFSI) occurred at 387 and 391 °C, respectively. The former remained a residue of about 9.4% at about 487 °C and the latter had about 7.2% residue at about 488.5 °C. An examination of these results reveals that the two samples have very close thermal decomposition data with anion exchange of TFSI$^-$ instead of Br$^-$. The thermal stability of the proposed polyelectrolytes is higher in the case of the TFSI anion than the Br anion, and this is in accordance with the DSC suggestions on the energy associated to their solid state arrangements.

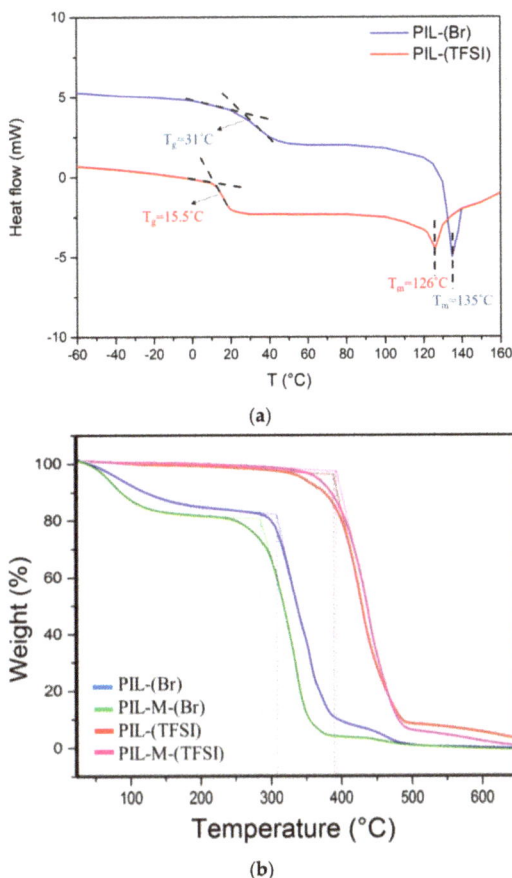

Figure 3. (**a**) Differential scanning calorimetric analysis of PIL-(TFSI) and PIL-(Br); (**b**) Thermal gravity analysis of PIL-(Br), PIL-M-(Br), PIL-(TFSI), and PIL-M-(TFSI).

3.4. Ionic Conductivity

Electrochemical impedance measurement can be used to calculate the ionic conductivity of the material. The ionic conductivity can be evaluated according to the following equation:

$$\sigma = \frac{l}{R_b \times A} \tag{1}$$

where σ is the conductivity in S·cm^{-1}, l is the thickness of the sample in cm, R_b is the resistance of the material in Ω, and A is the area of the sample in cm^2. Figure 4 shows the electrochemical impedance of PIL-M-(Br) and PIL-M-(TFSI). The value of R_b was determined by the intersection point on the *x*-axis of alternating current (AC) impedance spectroscopy. Based on Equation (1), the values of ionic conductivity were 5.52×10^{-4} and 3.2×10^{-4} S·cm^{-1} for PIL-M-(Br) and PIL-M-(TFSI), respectively, at 25 °C.

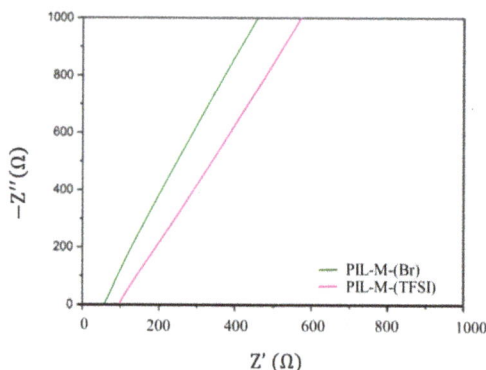

Figure 4. Alternating current (AC) impedance of PIL-IL blends.

3.5. Cyclic Voltammogram

Figure 5 shows cyclic voltammograms (CVs) of MBIB, PIL-M-(Br), MBIT, and PIL-M-(TFSI)-based supercapacitors. It is obvious that the shape of these CV curves is close to the rectangle, which is responsible for the characteristics of ideal electrical double layer capacitance behavior. Also note that the CV area decreases with the decreasing scan rate without a noticeable change in shape, indicating that the ions in the electrolyte can be rapidly migrated between the electrode interface and electrolyte [16]. The specific capacitance (C_s) of the electrodes can be calculated from the CV curves according to the following equation:

$$C_s = \frac{\int I dV}{V m \Delta V} \tag{2}$$

where m is the mass of the electroactive material (g), v is the scan rate (V·s^{-1}), ΔV is the potential window (V), and the integrated area under the CV curve I is the response current (A). According to Equation (1), the integrated area of the CV curve is proportional to the specific capacitance. Figure 6 shows the area enclosed by the CV curves of MBIB and PIL-M-(Br) using different scan rates. There is a small difference in specific capacitance in both electrolytes of MBIB and PIL-M-(Br) at different scan rates, implying that the capacitance values in both electrolytes of MBIB and PIL-M-(Br) are very close. This suggests that the solid-state PIL-M-(Br) electrolyte is suitable for use in a supercapacitor. However, the CV curve area of MBIB is higher than that of PIL-M-(Br) at high speed scan, indicating that the ion transport of MBIB is faster than PIL-M-(Br) because the ion mobility of liquid electrolyte is higher than that of solid electrolytes [16]. The CVs of MBIT and PIL-M-(TFSI) were similar to that of MBIB (Br) and PIL-M-(Br). In Figure 6, the CV area of MBIT is greater than that of PIL-M-(TFSI) at same scan rate, indicating that the ion transport in MBIT is significantly higher than that of PIL-M-(TFSI) because the ion mobility of the liquid electrolyte is higher than that of the solid electrolyte [16]. Especially, a larger anion size, such as TFSI$^-$, has more significant effect on the ion mobility. As compared to the CVs of PIL-M-(Br) and PIL-M-(TFSI) at a scan rate of 100 mV·s^{-1}, the latter has a larger potential window (3 V) than the former (2 V). However, the integrated area of the former is larger than that of the latter. This is attributed to the difference in ion conductivity between the two electrolytes, resulting from better ion transport in PIL-M-(Br) than in PIL-M-(TFSI).

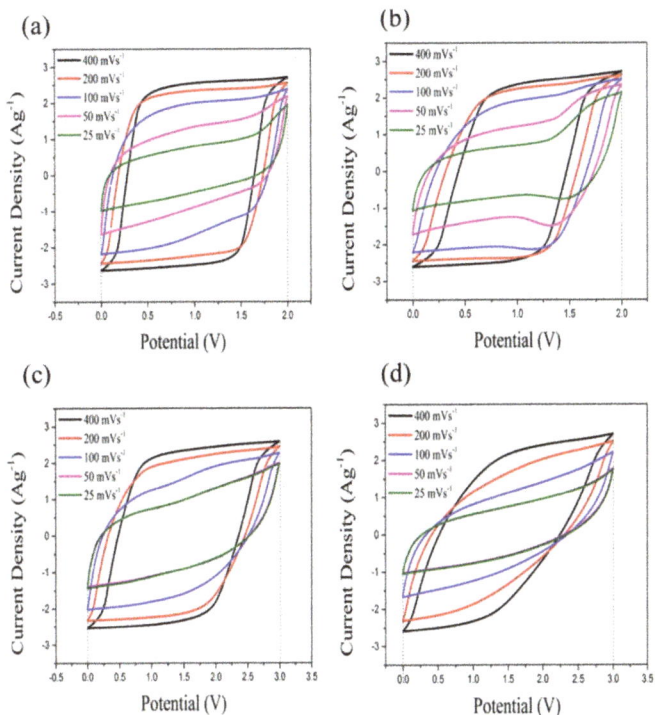

Figure 5. Cyclic voltammograms (CVs) of (**a**) MBIB; (**b**) PIL-M-(Br); (**c**) MBIT; and (**d**) PIL-M-(TFSI)-based supercapacitors at different scan rates.

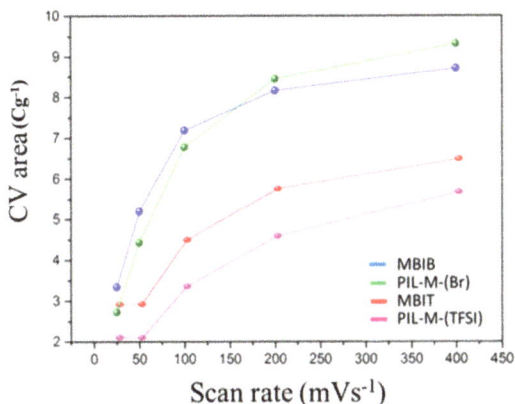

Figure 6. Plot of CV enclosed area against scan rate for ILs and PIL-IL blends.

3.6. Electrochemical Impedance Spectroscopy (EIS)

Nyquist plots of MBIB, PIL-M-(Br), MBIT, and PIL-M-(TFSI)-based supercapacitors are shown in Figure 7. The equivalent series resistance (ESR) and charge transport resistance (R_{ct}) are estimated in the high frequency region. The intersection of the high-frequency region with the real *x*-axis of the

Nyquist plot was used to estimate the equivalent series resistance (ESR) of the electrode. The ESR values of MBIB and PIL-M-(Br) are 13 and 16.2 Ω, respectively, due to the fact that the ionic conductivity of the liquid electrolyte (MBIB) is higher than that of the solid electrolyte (PIL-M-(Br)). The ESR values of MBIT and PIL-M-(TFSI$^-$) are 8.1 and 19.0 Ω, respectively. The charge transfer resistance (R_{ct}) at the contact interface between the electrode and electrolyte solution forms the charge transfer limiting process corresponding to the high-frequency arc [1]. Both values of R_{ct} are very small in MBIB- and PIL-M-(Br)-based supercapacitors, corresponding to the charge transfer impedance between the interface of the electrode and electrolyte, because there are almost no electrochemical reactions involved in the charge transfer in the electric double layer capacitors, resulting in very small R_{ct}. Also, both values of R_{ct} are very small in the MBIT- and PIL-M-(TFSI)-based supercapacitors. On the other hand, the straight line in the low-frequency range corresponds to the Warburg impedance, which is caused by the frequency dependence of ion diffusion and transport from the electrolyte to the electrode surfaces [1]. The MBIB and PIL-M-(Br) electrolytes exhibit the characteristics of ideal electric double layer capacitance, reflecting rapid ion diffusion in the electrolyte and adsorption onto the electrode surface. As compared to PIL-M-(Br), MBIB has a large slope in the straight line at low frequency, implying that the ions diffused in MBIB have less resistance than in PIL-M-(Br), mainly because the mass transfer in the liquid phase is easier than in the solid phase. Figure 8 shows the Bode plots of the MBIB and PIL-M-(Br)-based supercapacitors. At a phase angle of 45°, the response frequencies of MBIB and PIL-M-(Br) are 0.3 and 0.0552 Hz, whereas the τ_0 values are 3.33 and 18.115 s, respectively, indicating that the MBIB has a higher power density than PIL-M-(Br). In addition, the former has a faster charge–discharge speed because the ionic conductivity of the liquid electrolyte is higher than that of the solid electrolyte, resulting in the charge transferring more quickly between electrolyte and electrode. In MBIT and PIL-M-(TFSI) systems, the response frequencies at a phase angle of 45° are the same value of 0.098 Hz and the τ_0 value is the same value of 10.2 s, indicating that MBIT and PIL-M-(TFSI) have the same charge–discharge capacity under ideal capacitance conditions. Also note in Figure 7, that the slope of PIL-M-(Br) is larger than that of PIL-M-(TFSI) in the low frequency range, implying that the former has less resistant ion diffusion than the latter. This is attributed to the fact that the ionic conductivity of PIL-M-(Br) is greater than that of PIL-M-(TFSI). Table 3 summarizes the data obtained by the EIS analysis of PIL-M-(Br) and PIL-M-(TFSI)-based capacitors. An examination of Table 3 reveals that the ESR values of PIL-M-(Br) are lower than those of PIL-M-(TFSI). This reflects the fact that the former has lower solution resistance than the latter. In addition, the τ_0 values of PIL-M-(Br) are higher than those of PIL-M-(TFSI), suggesting that the latter capacitor has a higher charge–discharge rate than the former in the ideal state.

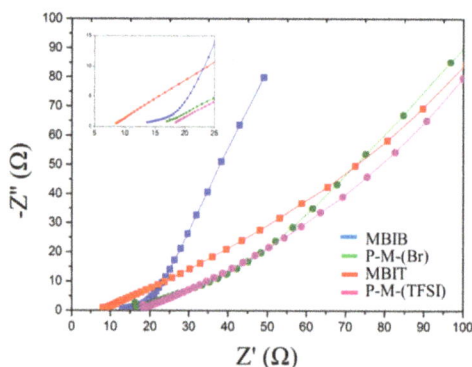

Figure 7. Nyquist plots of IL and PIL-IL blend-based supercapacitors.

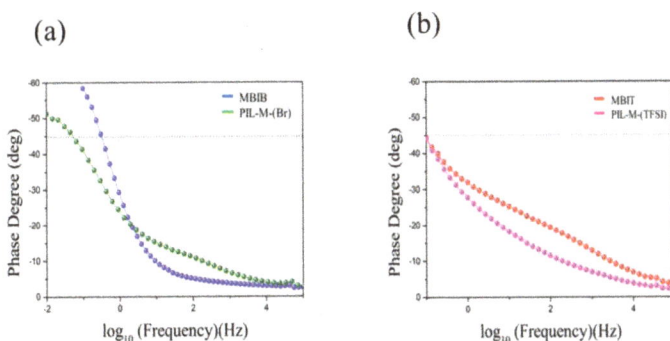

Figure 8. Bode plot of (**a**) MBIB and PIL-M-(Br) and (**b**) MBIT and PIL-M-(TFSI)-based supercapacitors.

Table 3. EIS data of PIL-M-(Br)- and PIL-M-(TFSI)-based supercapacitors.

Substance	ESR (Ω)	τ_0
PIL-M-(Br)	16.2	18.1
PIL-M-(TFSI)	19.0	10.2

3.7. Galvanic Charge–Discharge (GCD)

Figure 9 shows the galvanostatic charge–discharge curves at current densities of 10, 5, 2.5, 2 and 1 A·g^{-1} in the potential window of 0–2 V for MBIB- and PIL-M-(Br)-based supercapacitors. These charge–discharge curves have a symmetrical triangular shape, indicating that the ideal characteristics of the electrical double-layer capacitor (EDLC) provide these electrodes with excellent electrochemical reversibility [2]. However, different charge and discharge current densities did not change the line shape, demonstrating fast charge–discharge kinetic behavior [17]. Figure 10 shows the voltage (IR drop) at different charge–discharge current densities, revealing that the voltage increases with an increasing charge–discharge current density. This arises from the fact that the charge diffusion rate cannot match the rapid increase in current densities [18,19]. In addition, the slope of the straight line represents the bulk resistance. The bulk resistance of the MBIB-based capacitor is less than that of the PIL-M-(Br)-based capacitor, because the ionic conductivity of MBIB is greater than that of PIL-M-(Br). The specific capacitance values are calculated from the results of the constant current charge–discharge. Figure 11 shows the specific capacitance values of MBIB- and PIL-M-(Br)-based capacitors at different current densities. The specific capacitance of both systems decreases with an increasing charge–discharge current density. This is because the increase of the current density causes the increase of ion-diffusion resistance in the electrode material, leading to the decrease of the ion diffusion rate in the electrolyte, thereby lowering the capacitance. The specific capacitance of MBIB is slightly larger by about 10% than that of PIL-M-(Br). One explanation for this result is that MBIB has a higher ionic conductivity than PIL-M-(Br), resulting in an MBIB solution resistance smaller than PIL-M-(Br). Moreover, the mass transfer rate of MBIB is faster than that of PIL-M-(Br), which makes MBIB ions easier to adsorb and diffuse on the electrode material. The potential window (3 V) of PIL-M-(TFSI) is larger than that (2 V) of PIL-M-(Br). In addition, the specific capacitance was plotted against the current density as shown in Figure 11, revealing that the specific capacitance of the PIL-M-(Br)-based capacitor is higher than that of the PIL-M-(TFSI)-based capacitor. These results can be supported by a lower IR drop of PIL-M-(Br$^-$ system as compared PIL-M-(TFSI) (Figure 11), suggesting that the ionic conductivity of PIL-M-(Br) is greater than that of PIL-M-(TFSI).

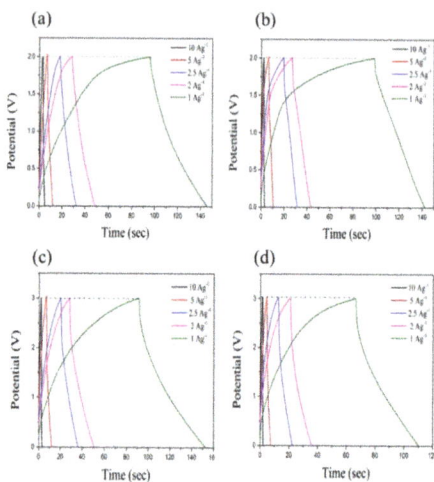

Figure 9. Galvanic charge–discharge curves of (**a**) MBIB; (**b**) PIL-M-(Br); (**c**) MBIT; and (**d**) PIL-M-(Br)-based supercapacitors at different current densities.

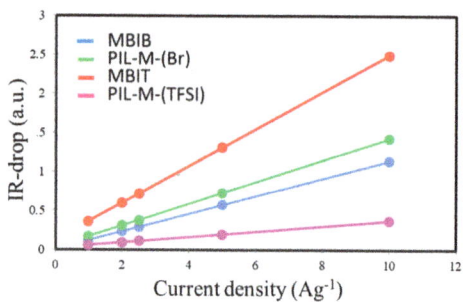

Figure 10. Plot of IR-drop against current density for ILs and PIL-IL blends.

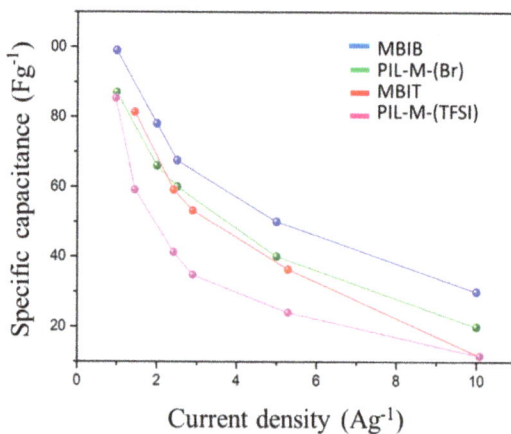

Figure 11. Plot of specific capacitance against current density for ILs and PIL-IL blends.

In Figure 12, specific energy density (Wh·kg^{-1}) and specific power density (W·kg^{-1}) can be calculated to obtain the Ragone plot. The MBIB-based supercapacitor has a slightly higher specific power density at the same specific energy density, and vice versa, indicating that the MBIB-capacitor can store a higher amount of electricity in a unit mass, whereas it can also store or release a higher power in a unit of time, but there are no large differences between the two systems of MBIB- and PIL-M-(Br)-based supercapacitors. Also note that the PIL-M-(Br)-based supercapacitor has a higher specific power density in relation to the PIL-M-(TFSI)$^-$-based supercapacitor at the same specific energy density. In contrast, the PIL-M-(TFSI)-based supercapacitor has a higher specific energy density at the same specific power density.

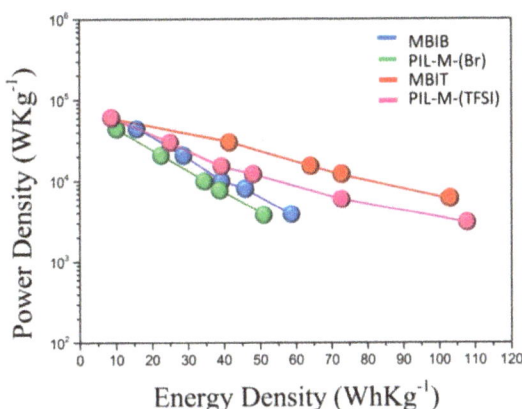

Figure 12. Ragone plot of MBIB and PIL-M-(Br$^-$)-based supercapacitors.

Figure 13 shows cycle life of the Coulombic efficiency in PIL-M-(Br)- and PIL-M-(TFSI)-based supercapacitors, revealing that the Coulomb efficiency maintains at 80% and 72%, respectively, after 2000 cycles. This is attributed to the fact that ions accumulate in the pores of the electrode material, reducing the redox process of the electrolyte. After repeated cycles, the electrolyte may deteriorate to decrease the ionic conductivity of the electrolyte with the increase in electrolyte resistance [17,20–23]. The PIL-M-(TFSI)-based supercapacitor has a similar tendency.

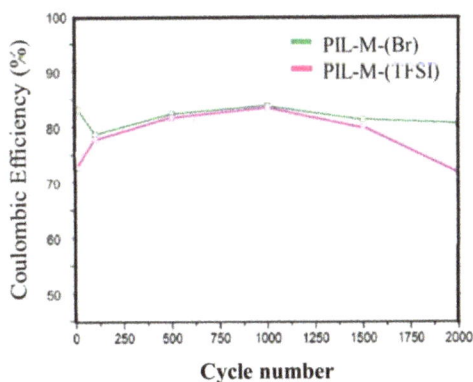

Figure 13. Cycle life of PIL-Il blend-based supercapacitors.

4. Conclusions

The ionic liquid polymer blends PIL-M-(Br) and PIL-M-(TFSI) had thermal stability with higher decomposition temperatures of 305 and 391 °C, respectively. The PIL-M-(Br) and PIL-M-(TFSI) had high ionic conductivity of 0.55 and 0.32 mS·cm^{-1}, respectively, which were very suitable for use in carbon-based symmetric supercapacitors, with the potential windows of 2 V and 3 V with respect to PIL-M-(Br) and PIL-M-(TFSI), respectively. Carbon-based symmetric supercapacitors involving the respective electrolytes PIL-M-(Br) and PIL-M-(TFSI) had a high power density and energy density of 59.9 and 40 kW·kg^{-1}, and 48 and 107 Wh·kg^{-1}, respectively. The PIL-M-(Br)-based supercapacitor had a higher specific capacitance than PIL-M-(TFSI) with specific capacitances of 87 and 58 F·g^{-1}, respectively, at a charge–discharge current density of 1 A·g^{-1}. In contrast, the values of the MBIB and MBIT ionic liquids were 99 and 82 F·g^{-1}, respectively. After 2000 cycles of charge–discharge testing, the efficiency decay of solid state PIL-M-(Br) and PIL-M-(TFSI)-based supercapacitors was 12 and 29%, respectively. The time constant (τ_o) of PIL-M-(Br) is longer than that of PIL-M-(TFSI), suggesting that the latter has a better charge–discharge rate under ideal capacitors.

Acknowledgments: Authors greatly acknowledged financial support from the Ministry of Science and Technology in Taiwan (MOST 106-2221-E-390-025 and MOST 106-2622-E-390-001-CC2).

Author Contributions: Po-Hsin Wang performed the experiments; Chien-Hsin Yang, Wen-Churng Lin and Tzong-Liu Wang conceived and designed the experiments; Po-Hsin Wang, Hung-Yin Lin, Mei-Hwa Lee and Chien-Hsin Yang analyzed the data; Hung-Yin Lin, Chien-Hsin Yang and Tzong-Liu Wang contributed reagents/materials/analysis tools; Chien-Hsin Yang, Po-Hsin Wang, Wen-Churng Lin, Hung-Yin Lin, Mei-Hwa Lee and Tzong-Liu Wang wrote the paper.

Conflicts of Interest: The authors declare no conflict of interest. The founding sponsors had no role in the design of the study; in the collection, analyses, or interpretation of data; in the writing of the manuscript, and in the decision to publish the results.

References

1. Burke, A. Ultracapacitors: Why, how, and where is the technology. *J. Power Sources* **2000**, *91*, 37–50. [CrossRef]
2. Simon, P.; Gogotsi, Y.; Dunn, B. Where do batteries end and supercapacitors begin? *Mater. Sci.* **2014**, *343*, 1210–1211. [CrossRef]
3. Schneuwly, A.; Gallay, R. Properties and applications of supercapacitors from the state-of-the-art to future trends. *Proc. PCIM* **2000**, 1–10.
4. Béguin, F.; Frąckowiak, E. *Supercapacitors Materials, Systems, and Applications*; John Wiley & Sons: Weinhein, Germany, 2013.
5. Kötza, R.; Carlenb, M. Principle and applications of electrochemical capacitors. *Electrochim. Acta* **2000**, *45*, 2483–2498. [CrossRef]
6. Ruiz, V.; Santamaría, R.; Granda, M.; Blanco, C. Long-term cycling of carbon-based supercapacitors in aqueous media. *Electrochim. Acta* **2009**, *54*, 4481–4486. [CrossRef]
7. Armand, M.; Endres, F.; MacFarlane, D.R.; Ohno, H.; Scrosati, B. Ionic-liquid materials for the electrochemical challenges of the future. *Nat. Mater.* **2009**, *8*, 621–629. [CrossRef]
8. Yamazaki, S.; Ito, T.; Yamagata, M.; Ishikawa, M. Non-aqueous electrochemical capacitor utilizing electrolytic redox reactions of bromide species in ionic liquid. *Electrochim. Acta* **2012**, *86*, 294–297. [CrossRef]
9. Brachet, M.; Brousse, T.; Bideaua, J.L. All Solid-state symmetrical activated carbon electrochemical double layer capacitors designed with ionogel electrolyte. *ECS Electrochem. Lett.* **2014**, *3*, 112–115. [CrossRef]
10. Ahn, Y.K.; Kim, B.; Ko, J.; You, D.J.; Yin, Z.; Kim, H.; Shin, D.; Cho, S.; Yoo, J.; Kim, Y.S. All solid state flexible supercapacitors operating at 4 V with a cross-linked polymer–ionic liquid electrolyte. *J. Mater. Chem. A* **2016**, *4*, 4386–4439. [CrossRef]
11. Tiruye, G.A.; Munoz-Torrero, D.; Palma, J.; Anderson, M.; Marcilla, R. All-solid state supercapacitors operating at 3.5 V by using ionic liquid based polymer electrolytes. *J. Power Sources* **2016**, *279*, 472–480. [CrossRef]
12. Tehfe, M.A.; Louradour, F.; Lalevée, J.; Fouassier, J.P. Photopolymerization reactions: On the way to a green and sustainable chemistry. *Appl. Sci.* **2013**, *3*, 490–514. [CrossRef]

13. Bowers, D.I.; Maddams, W.F. Radiation curing of polymeric materials. *Anal. Chim.* **1990**, *62*, 668.
14. Chen, S.M.; Wang, T.L.; Chang, P.Y.; Yang, C.H.; Lee, Y.C. Poly(ionic liquid) prepared by photopolymerization of ionic liquid monomers as quasi-solid-state electrolytes for dye-sensitized solar cells. *React. Funct. Polym.* **2016**, *108*, 103–112. [CrossRef]
15. Yuan, J.; Antonietti, M. Poly(ionic liquid) latexes prepared by dispersion polymerization of ionic liquid monomers. *Macromolecules* **2011**, *44*, 744–750. [CrossRef]
16. Yagci, Y.; Yilmaz, F.; Kiralp, S.; Toppare, L. Photoinduced polymerization of thiophene using iodonium salt. *Macromol. Chem. Phys.* **2005**, *206*, 1180–1182. [CrossRef]
17. Ryu, K.S.; Lee, Y.; Han, K.S.; Parka, Y.J.; Kang, M.G.; Park, N.G.; Chang, S.H. Electrochemical supercapacitor based on polyaniline doped with lithium salt and active carbon electrodes. *Solid State Ionics* **2004**, *175*, 765–768. [CrossRef]
18. Jang, H.S.; Raj, C.J.; Lee, W.G.; Kim, B.C.; Yu, K.H. Enhanced supercapacitive performances of functionalized activated carbon in novel gel polymer electrolytes with ionic liquid redox-mediated poly(vinyl alcohol)/phosphoric acid. *RSC Adv.* **2016**, *6*, 75376–75383. [CrossRef]
19. Mecerreyes, D. Polymeric ionic liquids: Broadening the properties and applications of polyelectrolytes. *Prog. Polym. Sci.* **2011**, *36*, 1629–1648. [CrossRef]
20. Chang, P.Y.; Yang, C.H. Photopolymerization of electroactive film applied to full polymer electrochromic device. *Express Polym. Lett.* **2017**, *11*, 176–186. [CrossRef]
21. Liew, C.W.; Ramesh, S.; Arof, A.K. Good prospect of ionic liquid based-poly(vinyl alcohol) polymer electrolytes for supercapacitors with excellent electrical, electrochemical and thermal properties. *Int. J. Hydrogen Energy* **2014**, *39*, 2953–2963. [CrossRef]
22. Liew, C.W.; Ramesh, S.; Arof, A.K. Characterization of ionic liquid added poly (vinyl alcohol)-based proton conducting polymer electrolytes and electrochemical studies on the supercapacitors. *Int. J. Hydrogen Energy* **2015**, *40*, 852–862. [CrossRef]
23. Jannasch, P. Ion conducting electrolytes based on aggregating comblike poly(propylene oxide). *Polymer* **2001**, *42*, 8629–8635. [CrossRef]

nanomaterials

MDPI

Article

Enhanced Supercapacitor Performance Using Electropolymerization of Self-Doped Polyaniline on Carbon Film

Po-Hsin Wang [1], Tzong-Liu Wang [1], Wen-Churng Lin [2], Hung-Yin Lin [1], Mei-Hwa Lee [3] and Chien-Hsin Yang [1,*]

[1] Department of Chemical and Materials Engineering, National University of Kaohsiung, Kaohsiung 81148, Taiwan; p5802341@gmail.com (P.-H.W.); tlwang@nuk.edu.tw (T.-L.W.); linhy@nuk.edu.tw (H.-Y.L.)

[2] Department of Environmental Engineering, Kun Shan University, Tainan71070, Taiwan; linwc@mail.ksu.edu.tw

[3] Department of Materials Science and Engineering, I-Shou University, Kaohsiung 84001, Taiwan; mhlee@isu.edu.tw

* Correspondence: yangch@nuk.edu.tw; Tel.: +886-7-5919420

Received: 8 February 2018; Accepted: 27 March 2018; Published: 31 March 2018

Abstract: In this work, we electrochemically deposited self-doped polyanilines (SPANI) on the surface of carbon-nanoparticle (CNP) film, enhancing the superficial faradic reactions in supercapacitors and thus improving their performance. SPANI was electrodeposited on the CNP-film employing electropolymerization of aniline (AN) and o-aminobenzene sulfonic acid (SAN) comonomers in solution. Here, SAN acts in dual roles of a self-doped monomer while it also provides an acidic environment which is suitable for electropolymerization. The performance of SPANI−CNP-based supercapacitors significantly depends upon the mole ratio of AN/SAN. Supercapacitor performance was investigated by using cyclic voltammetry (CV), galvanostatic charge and discharge (GCD), and electrochemical impedance spectroscopy (EIS). The optimal performance of SPANI−CNP-based supercapacitor exists at AN/SAN ratio of 1.0, having the specific capacitance of 273.3 Fg^{-1} at the charging current density of 0.5 Ag^{-1}.

Keywords: supercapacitor; self-doped polyaniline; carbon-nanoparticle film; electropolymerization; cyclic voltammetry

---

1. Introduction

Energy storage systems are important nowadays because of the growing requirement for high power in charging and discharging, and thus the development of supercapacitors has attracted extensive attention recently. Supercapacitors have been categorized into electrochemical double layer capacitors (EDLCs) and pseudocapacitors according to their mechanism of charge storage [1]. EDLCs store charge employing electrical double-layer electrostatic force in an electrochemical double layer (Helmholtz Layer) [2]. The main materials used in EDLC are carbon-based materials, such as graphene, carbon nanotubes, carbon black, etc. In pseudocapacitors, charge is stored via fast and reversible faradic reaction, which is based on electroactive materials with several oxidation and reduction states (e.g., metal oxides and conducting polymer). In general, the capacitance of pseudocapacitors are higher than EDLCs [3,4]. On the other hand, EDLCs use very stable carbons to increase voltage (V) [5,6].

Conducting polymers, such as polyaniline (PANI) which is a potential material for supercapacitors due to its excellent electrochemical activity, ease of synthesis, environmental stability, and high specific capacitance [7], have been used for supercapacitors. Based on previous studies, the nanosized materials can improve supercapacitor electrodes by increasing the area of the electrode–electrolyte interface and decreasing the cation diffusion length within the active material [8–10]. Electrically conductive nanosized PANI is combined with porous carbon, carbon nanotubes, or graphene as a nanostructured template that not only improves the charge/discharge cycle of PANI but also promotes the specific capacitance to 233–1220 Fg^{-1} [8,11–14].

Self-doped polyanilines (SPANI) nanofibers have been synthesized by the molecular self-assembly process using aniline (AN) and o-aminobenzenesulphonic acid (SAN) without extra addition of inorganic acids. SAN acted as a self-doping monomer, surfactant, and template for PANI nanofibers due to its hydrophilic group ($-SO_3H$). Because SAN acted as a self-doping monomer and was simultaneously linked in the polymer chains, the SAN didn't need removal after polymerization. We tried to employ this method devoid of extra addition of inorganic acids, which is significantly different from the alternative template methods and external doping systems [15,16].

In this work, SPANIs with different AN/SAN ratios were deposited on the surface of carbon nanoparticle (CNP) film by electropolymerization forming SPANI−CNP composite electrodes. The SPANI−CNP composites are characterized by their morphologies, structure, and composition through examination by scanning electron microscopy (SEM), X-ray diffraction (XRD), Fourier transfer infrared spectrometer (FTIR), and X-ray photoelectron spectroscopy (XPS). The effects of AN/SAN ratio in SPANI−CNP composite electrodes on the performance of supercapacitors was investigated in three-electrode type and symmetric cells, respectively. The supercapacitor performance of the SPANI−CNP composite electrodes with different AN/SAN ratios was studied by cyclic voltammetry (CV), galvanostatic charge–discharge (GCD), and electrochemical impedance spectroscopy (EIS). The assembled SPANI−CNP-based supercapacitors demonstrate the optimal performance, having SPANI film at the AN/SAN ratio of 1.0 which provides channels for rapid transportation of electrolyte ions coupled with extra pseudocapacitive capabililty.

2. Experimental Section

2.1. Materials

Aniline (Aldrich, Milwaukee, WI, USA) was distilled under reduced pressure. o-Aminobenzenesulfonic acid (SAN, Fluka, Buchs, Switzerland) was recrystallized two times in distilled water. Carbon nanoparticles (~13 nm, Colour Black FW200, UniRegion Bio-Tech, Taip, Taiwan), poly(vinylidene fluoride) (PVDF, Solf®PVDF 6020, Solvay, Bruxelles, Belgium), 1-methyl-2-pyrrolidone (NMP, Riedel-de Haën, Seelze, Germany), graphite foil (Hongye Vacuum Technology Co., Ltd., Xinzhu, Taiwan), and sulfuric acid (Fluka, Buchs, Switzerland) were used as received.

2.2. Preparation of CNP Substrates

The carbon nanoparticle (CNP) electrodes were prepared by the following procedures. Graphite foil as the current collector was polished by sandpaper and washed by 0.05 M H_2SO_4, then dried in a vacuum oven at 120 °C for 2 h. Then, carbon nanoparticles and PVDF were mixed in a mass ratio of 9:1 by Agate mortar and dispersed in the NMP solvent, grinding until the paste was homogeneous. The paste was dropped onto a graphite foil (2 cm × 1 cm) and dried at 70 °C for 16 h in a vacuum oven. The loading mass of CNP was about 0.5 mg·cm^{-2}.

2.3. Preparation of SPANI−CNP Composite Electrodes

Monomers of AN and SAN was dissolved in 50 mL of distilled water with total concentration of 0.1 M at various mole ratio of AN/SAN from 0.5 to 4. SPANI was electrodeposited on CNP substrates forming SPANI−CNP composite electrodes using cyclic voltammetry in a three-electrode cell equipped with Pt counter electrode and reference electrode of Ag/AgCl. The potentials swept from −0.2 to

0.8 V (vs. Ag/AgCl) scanning ca. 40 cycles at the scan rate of 25 mV·s^{-1}. The deposited SPANI–CNP composite electrodes were dried in vacuum oven at 60 °C for 16 h.

2.4. Characterization Techniques

The surface morphology of the SPANI–CNP composites was identified by field-emission scanning electron microscopy (FE-SEM, JSM-6700F, JEOL Ltd., Tokyo, Japan). Infrared spectra were recorded on an FTIR spectrometer (Agilent Technologies, Cary 630, Santa Clara, CA, USA) to check the functional groups on polymer films. The structural and chemical composition were determined by X-ray photoelectron spectroscopy (XPS) using a Theta Probe AR-XPS (Theta 300 Version, Thermo Fisher Scientific, Paisley, UK) which scanned the surface of the sample with a monochromatic X-Ray beam source of 1486.6 eV (aluminum anode) and 15 kV.

2.5. Electrochemical Characterization

Electrochemical characterization of SPANI–CNP composite electrodes in the respective three-electrode cells and symmetric cells were analyzed using cyclic voltammetry (CV), galvanostatic charge and discharge (GCD), and electrochemical impedance spectroscopy (EIS) on an AUTOLAB PGSTAT302N electrochemical work station (Metrohm Autolab, Utrecht, Netherland). 1 N H$_2$SO$_4$ was quoted as the electrolyte solution. Three-electrode cells were equipped with the SPANI–CNP working electrode, platinum plate counter electrode, and reference electrode of Ag/AgCl electrode, respectively. The cyclic voltammetry (CV) data were recorded with scan rates of 5 and 10 mV·s^{-1} over the potential window of −0.2–0.8 V in a three-electrode cell. In the symmetric cell, the CV data were obtained at various scan rates of 10, 25, 50, 75, 100, and 125 mV·s^{-1} over the potential window of 0–0.8 V. The galvanostatic charge–discharge (GCD) curves were obtained over the potential range of 0–0.6 V at various current densities of 0.5, 1, 1.5, 2, 2.5, 5, and 10 A·g^{-1} in symmetric devices. A current density of 1 and 5 A·g^{-1} was used in three-electrode experimental cells over the potential range of 0–0.8 V (vs. Ag/AgCl). The electrochemical impedance spectroscopy (EIS) measurements were carried out between the frequency range of 0.1 Hz to 100 kHz at an amplitude of 10 mV.

3. Results and Discussion

3.1. Cyclic Voltammograms of SPANI Electropolymerization on CNP Substrate

Electropolymerization of polyaniline needs attachment and chain extension between active intermediates of diradical dications [17] generated from the monomers and propagating oligomer chains. The diradical dication is an energetic electrophile, so polymerization is generally performed in the presence of a strong acid (e.g., 1 M HCl, pH = −0.15) to deposit polyaniline on the electrode. Because the SAN molecule has a sulfonic acid group (–SO$_3$H) in the ortho-position on the benzene ring, SAN molecules can generate the diradical dications (the bipolaronic form of pernigraniline, presents peak at ~800 mV in pH = −0.15) [18] through self-doping amino groups (–NH$_2$ on AN and SAN molecules). This species is an energetic electrophile, extracting an electron from AN and SAN and becoming a radical cation which bonds with another radical. The increase in the intensity of peak at ~250 mV (in pH = −0.15) [18] indicates that these radical cations undergo bonding to form the polymer deposited on the substrate. Alternatively, we designed that the sulfonate-containing SAN mixed with AN as a comonomer to make a weakly acidic aqueous solution (pH = 5.0) without adding inorganic acid. Figure 1 shows the cyclic voltammogramms (CVs) of AN-SAN electropolymerization on a CNP substrate. In this weak acidic solution, the current still gradually increases with increasing the cycling number of electropolymerization, due to the increase in deposition of the conductive polymer. As above, when adding 1 M HCl to AN-SAN comonomers (pH= −0.15), the corresponding CVs of electropolymerization presented two distinct redox couples around at ~250 mV and ~800 mV [18]. But in this work (pH = 5.0), the above two peaks of diradical dications and radical cation are merged together to become indistinguishable in this weakly acidic environment.

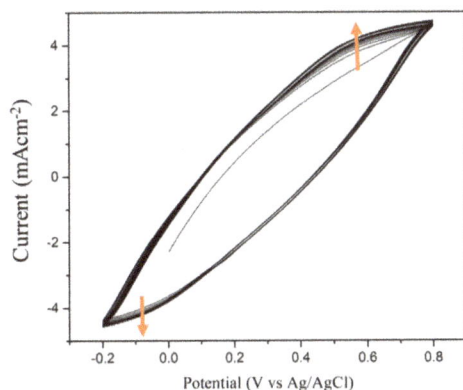

Figure 1. Cyclic voltammograms of self-doped polyanilines (SPANI) deposited on a carbon nanoparticle (CNP) substrate using electropolymerization with AN/SAN mole ratio of 1.0 (pH = 5.0) at a scan rate of 25 mV·s^{-1}. Yellow arrows represent the direction of current increase.

3.2. Material Analysis

Figure 2 shows the morphology of the CNP and SPANI−CNP composites. The CNP nanoparticles with the size of 20 nm can be seen in Figure 2a. The SPANI was assembled onto CNP film, thus enlarging the particle size to ~50 nm as shown in Figure 2a,b. As compared the size of CNP nanoparticles, the increase of particle size of SPANI−CNP composites arises from assembling the SPANI deposition onto CNP particles having SPANI polymers acting as a wrapping layer.

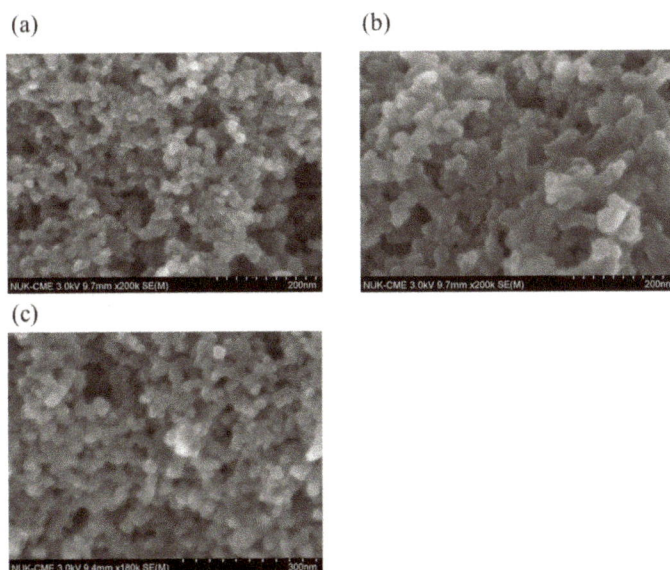

(a)

(b)

(c)

Figure 2. SEM images of (**a**) carbon nanoparticles; (**b**) SPANI−CNP composites with an AN/SAN ratio of 1.0; and (**c**) SPANI−CNP composites with an AN/SAN ratio of 4.0.

Figure 3 shows FTIR spectra of SPANI−CNP composites with AN/SAN ratio of 1 and 4, respectively. There are similar FTIR results in the two SPANI−CNP composites. The characteristic peaks of 1375 cm^{-1} and 1513 cm^{-1} are assigned to C=C stretching vibration of the quinoid ring and benzenoid ring, respectively. The bands at 1204 cm^{-1} and 1304 cm^{-1} are assigned to C–H stretching vibration with aromatic conjugation [16].

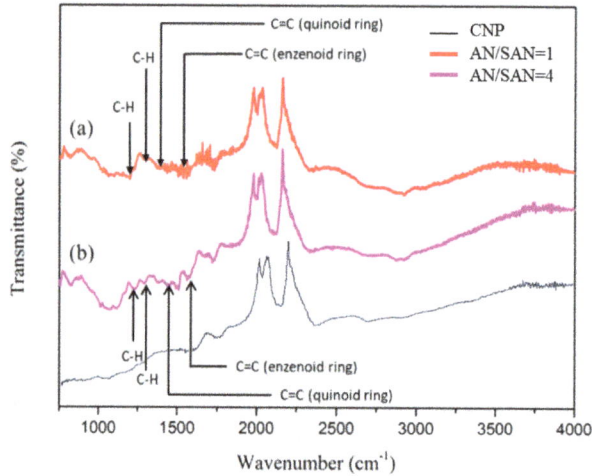

Figure 3. FTIR spectra of SPANI–CNP composites with AN/SAN ratio of (a) 1.0 and (b) 4.0.

In Figure 4, the atomic composition of the SPANI–CNP composites with different AN/SAN ratio is studied using X-ray photon spectroscopy (XPS), exhibiting the existence of C, N, O, and S elements. The N$_{1s}$ deconvolution of the SPANI–CNP composites with AN/SAN ratio of 1 and 4 are shown in Figure 5a,b. The binding energies of 399.2, 399.8, and 400.8 eV correspond to the imine (–N=), amine (–NH–), and polaron species (N$^+$), respectively [12]. The formation of polaron species is contributed by the nitrogen in the vicinity of H$^+$ cations when the polymers were self-doped with –SO$_3$H groups on the polymer chains. The formation of imine sites is due to the strong hydrogen bonding of –NH to oxygen atoms. Table 1 lists the results of N$_{1s}$ deconvolution and doping degree ([N$^+$]/[N]), revealing the doping degree of SPANI–CNP composites with AN/SAN ratio of 1 higher than that with AN/SAN ratio of 4. This is because the former has a higher content of SAN compared to the latter. Based on previous report, the electrical conductivity of SPANI is proportional to the doping degree [16]. As a result, it is conceivable that the conductivity of SPANI–CNP composites with AN/SAN ratio of 1 is higher than that with AN/SAN ratio of 4.

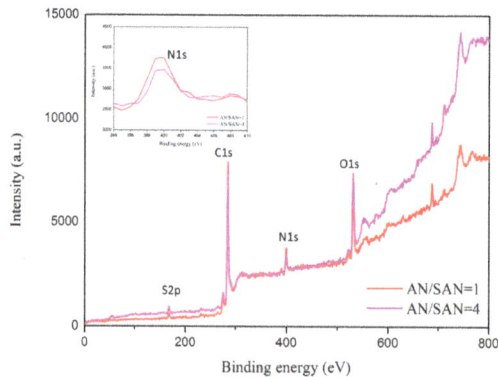

Figure 4. XPS spectra of SPANI–CNP composites with different AN/SAN ratios.

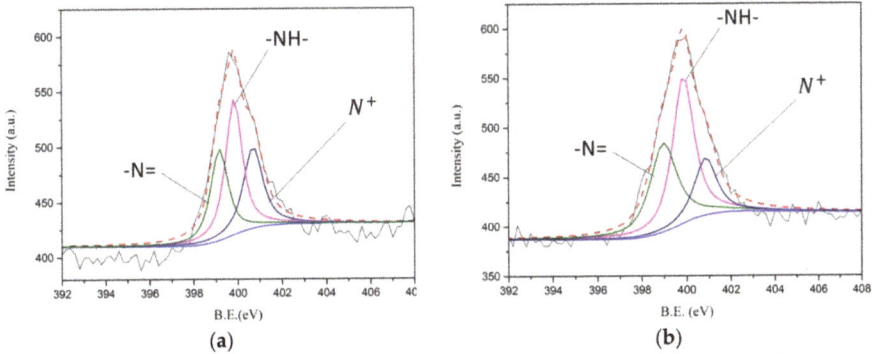

Figure 5. N_{1s} XPS core-level spectra of the SPANI−CNP with AN/SAN mole ratio of (a) 1 and (b) 4.

Table 1. Relative ratio (%) of different nitrogen components in SPANI–CB composites from N_{1s} XPS spectra.

AN/SAN	−N = (eV/%)	−NH− (eV/%)	N⁺ (eV/%)
1	399.2 (28.5%)	399.8 (39%)	400.8 (32.4%)
4	399 (31.6%)	399.8 (42.1%)	401 (26.2%)

3.3. Electrochemical Studies

Electrochemical techniques of cyclic voltammetry (CV), electrochemical impedance spectroscopy (EIS), and galvanostatic charge–discharge were employed to study the performance of CNP and SPANI−CNP composite-based electrodes in a three-electrode cell and a symmetric cell, respectively. The specific capacitance value (C_s) of CNP- and SPANI−CNP composite-based electrodes can be calculated from the CV curves according to equation [8]:

$$C_s = \frac{\int I dv}{v m \Delta V} \quad (1)$$

where m is the mass of the electroactive material (g), v is the sweep rate (V·s⁻¹), ΔV is the potential range (V), and the integrated area under the CV curve I is the response current (A). According to Equation (1), the integrated area of CV curve is proportional to specific capacitance of electrodes, so the specific capacitance of the CNP electrode and SPANI−CNP composite electrodes with AN/SAN ratios of 0.5, 1.0, 2.3, and 4.0 were compared. Figure 6a,b shows the cyclic voltammograms (CVs) in a three-electrode cell at sweep rate of 5 and 10 mV·s⁻¹ in 1 N H₂SO₄ solution. In these CVs, the peaks of oxidation and reduction peaks were observed from the characteristics of SPANI. Figure 6a,b also indicate that the SPANI−CNP composite electrodes showed a higher specific capacitance and electrochemical performance as compared CNP electrode. Moreover, the SPANI−CNP composite electrode with AN/SAN ratio of 1.0 possessed the highest specific capacitance than other AN/SAN compositions. These could be attributed to higher electrical conductivity of SPANI and its Faradic contribution.

Faradic contribution provided pseudocapacitance which was caused by reversible redox transition involving the exchange of protons (or cations) in the electrolyte (H₂SO₄) explained as follows.

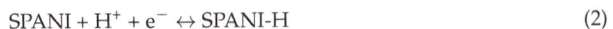

$$\text{SPANI} + \text{H}^+ + \text{e}^- \leftrightarrow \text{SPANI-H} \quad (2)$$

where SPANI:

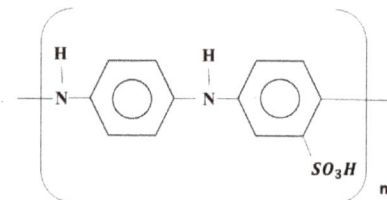

where H$^+$ denotes the protons in the electrolyte (H$_2$SO$_4$). Protons come from the sulfonic acids of SAN units in polymer chains. This implies that the formation of more complete polaronic structure of SPANI chains at this case. The above equation suggests that both protons are involved in the redox process. Here, SPANI promotes faster charge–discharge rates through the Faradaic reaction arising from doping–dedoping of the self-doped groups of sulfonic acid in the SAN units. These results further indicate that charge stored at the surface of the SPANI from the pseudocapacitance storage mechanism is an important factor in achieving high gravimetric energy density values.

(a) (b)

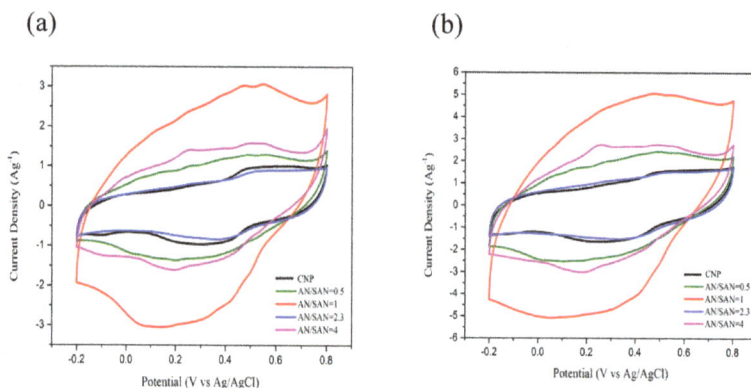

Figure 6. CV curves of SPANI−CNP composite electrodes with AN/SAN ratio at scan rates of (a) 5 mV·s^{-1} and (b) 10 mV·s^{-1}.

Figure 7a,b shows the CVs in a symmetric cell based on SPANI−CNP composite and CNP electrodes at sweep rate of 75 and 100 mV·s^{-1} in 1 N H$_2$SO$_4$ solution. The symmetric cell based on the SPANI−CNP composite electrode with an AN/SAN ratio of 1.0 yielded the largest area in the CVs. Figure 7c shows the CVs in a symmetric cell based on SPANI−CNP composite electrode supercapacitor with an AN/SAN ratio of 1.0 at scan rates of 10, 25, 50, 75, 100, and 125 mV·s^{-1}. These CVs have an almost rectangular shape from 10 mV·s^{-1} to 125 mV·s^{-1}, indicating fast transport of electrolyte ions and excellent capability. This represents that the charge transport on the surface of this electrode is highly efficient [9]. Figure 7d shows the CVs of the SPANI−CNP composite electrode with an AN/SAN ratio of 1.0 at various voltage windows, indicating that the shape of these CV curves remains almost identical in the potential range of 0.2 V to 1.0 V. This means that the SPANI−CNP composite electrode with an AN/SAN ratio of 1.0 has excellent capacitive characteristics and reversibility. The CVs exhibit a tail at the upper limit potential of 1.0 V, indicating the evolution of oxygen [9].

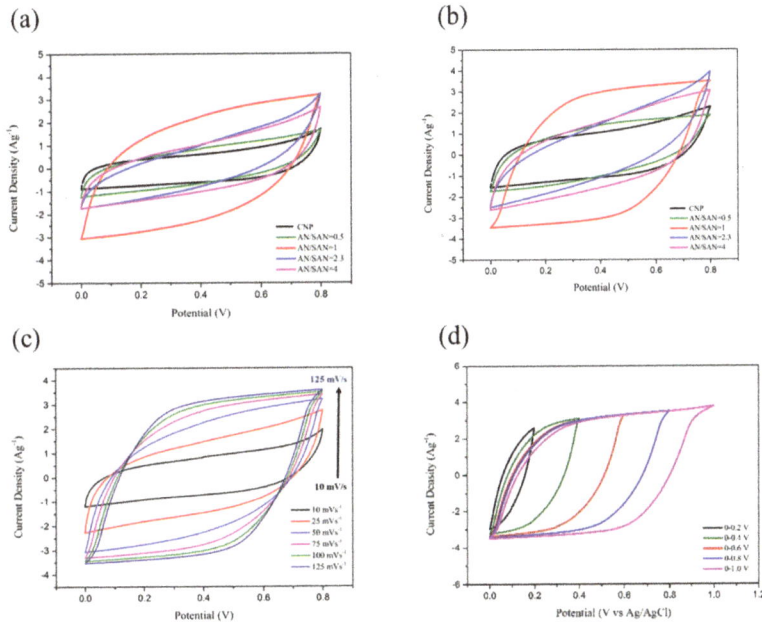

Figure 7. CV curves of SPANI−CNP composites with different AN/SAN ratios and CNP electrodes at scan rates of (**a**) 50 mV·s^{-1} and (**b**) 100 mV·s^{-1}; (**c**) CV curves of SPANI−CNP composites with different AN/SAN ratios and CNP electrodes at scan rates of 100 mV·s^{-1}; (**d**) CV curves of SPANI−CNP composite with AN/SAN ratio of 1.0 in the voltage-window range of 0–1.0 V.

3.4. Electrochemical Impedance Spectroscopy

The interface behavior of the electrode materials in supercapacitors is usually studied using EIS analysis [15]. The EIS experiment was performed in the frequency range of 0.1 to 100 kHz at an amplitude of 10 mV. In Figure 8, the Nyquist plot of CNP and SPANI−CNP composite electrodes showed a semicircular arc in the high-frequency range and a straight line in the low-frequency range. The arc of the high-frequency range indicates the charge transfer limiting process arising from the double-layer capacitance coupled with the charge transfer resistance (R_{ct}) at the contact interface between the electrode and electrolyte solution [10]. Figure 8 shows that the SPANI−CNP composite electrode with an AN/SAN ratio of 1.0 has the smallest diameter of the semicircular arc at the high-frequency region. The values of R_{ct} for the CNP and SPANI−CNP composite electrodes with AN/SAN ratios of 0.5, 1.0, 2.3, and 4.0 are 1.0, 1.89, 1.63, 4.0, and 1.97, respectively (refer to Table 2). The SPANI−CNP composite electrode with an AN/SAN ratio of 1.0 exhibited the lowest R_{ct} value among these SPANI−CNP composite electrodes, indicating that the SPANI composite with AN/SAN ratio of 1 has a relatively lower R_{ct} and better capacitive behavior with small diffusion resistance [10]. On the other hand, the interception on the x-axis at the high-frequency region was used to estimate the equivalent series resistance (ESR). The ESR values can be obtained for the CNP and SPANI−CNP composite electrodes with AN/SAN ratios of 4.0, 2.3, 1.0, and 0.5 are 2.83, 2.9, 4.1, 2.39, and 3.23 Ω, respectively (refer to Table 2). The SPANI−CNP composite electrode with an AN/SAN ratio of 1.0 exhibits the lowest ESR value. This situation could be attributed to the highest conductivity of SPANI with an AN/SAN ratio of 1.0 [16] and the incorporation of SPANI into the composite structure enhancing efficient transport of electrolyte ions to the surface of the composite electrodes. The SPANI polymer chains facilitate the attachment of H$^+$ ions for electrical neutrality. In the low-frequency region, the straight line corresponds to the Warburg resistance, which is caused by frequency dependence of ion diffusion and transport among the electrolyte to the electrode surfaces [10]. The almost vertical line of the SPANI−CNP composite electrode with an AN/SAN ratio of 1.0 in the low-frequency range indicates fast ion diffusion in the electrolyte and adsorption/desorption onto the electrode surface.

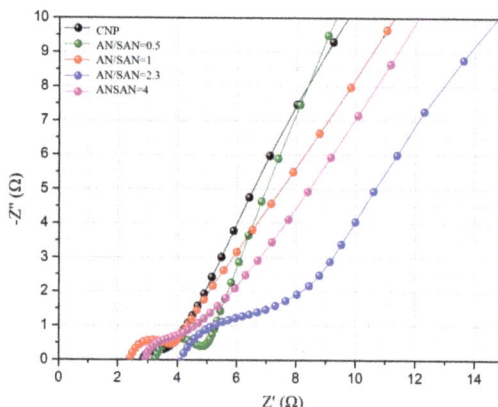

Figure 8. Nyquist plot of CNP and SPANI−CNP composite electrodes.

Table 2. ESR and R_{ct} values of CNP and SPANI−CNP composite electrodes.

Sample	ESR (Ω)	R_{ct} (Ω)
CNP	2.83	1.0
AN/SAN = 0.5	3.23	1.89
AN/SAN = 1	2.39	1.63
AN/SAN = 2.3	4.1	4.0
AN/SAN = 4	2.9	1.97

3.5. Galvanostatic Charge−Discharge

The specific capacitance of three-electrode and tow-electrode systems can be calculated from the galvanostatic charge–discharge curve at particular current densities by using Equations (3) and (4) [12], respectively:

$$C_s = \frac{I \Delta t}{m \Delta V} \tag{3}$$

$$C_s = \frac{4 I \Delta t}{m \Delta V} \tag{4}$$

where I is the discharge current (mA), m is the mass of total electroactive material (g), ΔV is the potential range of charge and discharge, and Δt is the discharge time (t). Figure 9 shows the galvanostatic charge and discharge curves of SPANI−CNP electrodes in a three-electrode system, where CNP and SPANI−CNP composites, platinum, and Ag/AgCl were used as the working electrode, the counter electrode, and the reference electrode, respectively, at a constant current density of 1 and 5 A·g^{-1} in 1 N H$_2$SO$_4$ electrolyte solution within a potential window of 0–0.8 V. The charge and discharge curves on CNP electrodes have a symmetric triangular shape, indicating the behavior of an electrical double-layer capacitor (EDLC) [11]. But, a deviation from a straight line was shown in the galvanostatic charge–discharge curves of the SPANI−CNP composite electrodes, indicating pseudocapacitive behavior [14]. The SPANI−CNP composite electrode with an AN/SAN ratio of 1.0 exhibited the highest specific capacitance in these devices.

Figure 10a shows the galvanostatic charge and discharge curves of SPANI−CNP electrodes in a symmetric cell at a constant current density of 1 A·g^{-1} in 1 N H$_2$SO$_4$ electrolyte solution within a potential window of 0–0.6 V. The specific capacitances of the CNP and the SPANI−CNP-based supercapacitors in the symmetric cells with AN/SAN ratios of 0.5, 1.0, 2.3, and 4.0 are 68, 53.3, 213.3, 102.6 and 96.6 F·g^{-1}, respectively, at current density of 1 A·g^{-1}. The SPANI−CNP composite with an AN/SAN ratio of 1.0 had the highest specific capacitance among these symmetric cells. Figure 10b shows the discharging rate of SPANI−CNP composite with AN/SAN ratio of 1 at different current density, revealing that the discharging time increased with decreasing of the discharging current density. Figure 10c demonstrates the rate capability of CNP and SPANI−CNP-based capacitors, indicating that the specific capacitance decreases with increasing current density and the specific capacitance of the SPANI−CNP-based capacitor is higher than that of CNP-based capacitor.

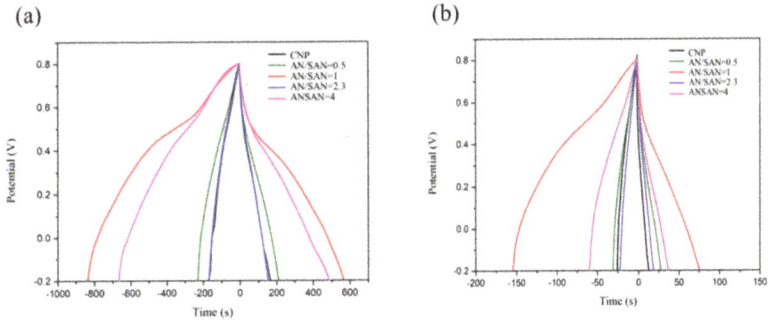

Figure 9. Galvanostatic charge and discharge curve of CNP and SPANI−CNP composite electrodes with AN/SAN ratio at current density of (**a**) 1 A·g^{-1} and (**b**) 5 A·g^{-1} in 1 N H$_2$SO$_4$.

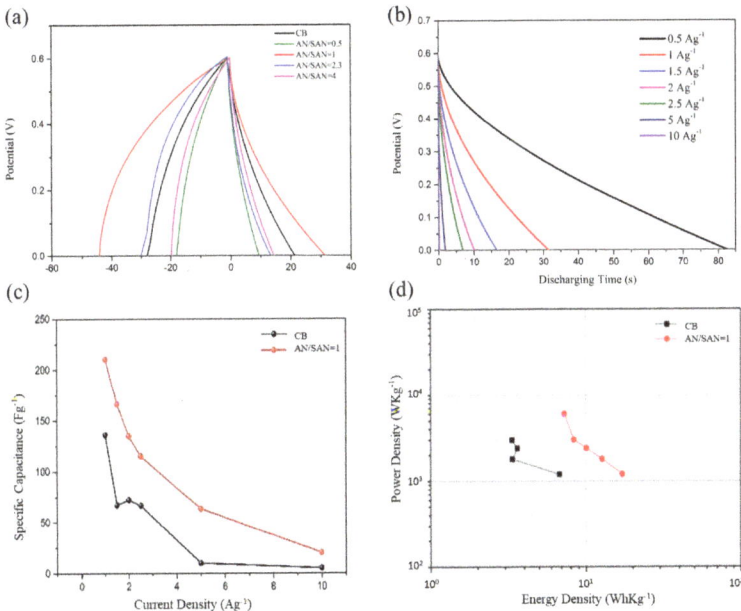

Figure 10. (**a**) Galvanostatic charge and discharge curves of CNP and SPANI−CNP composite electrodes with AN/SAN ratio at current density of 1 A·g^{-1}; (**b**) Discharging times of SPANI−CNP with AN/SAN = 1 composite at different current density; (**c**) Rate capability of CNP and SPANI−CNP composite electrodes; (**d**) Ragone Plot of CNP and SPANI−CNP-based supercapacitors.

Specific energy density (E, Wh·kg^{-1}) and specific power density (P, W·kg^{-1}) can be calculated using the following equations [14]:

$$E = \frac{1}{2}C_s V^2 \tag{5}$$

$$P = \frac{E}{t} \tag{6}$$

where C_s is the specific capacitance of the supercapacitors, V is the voltage range during the discharge process in galvanostatic charge and discharge curves, and t is the discharge time. Figure 10d shows the Ragone plot, indicating improvement in energy density (14.1 Wh/Kg) of SPANI−CNP-based capacitor with AN/SAN ratio of 1 as compared that (6.8 Wh/Kg) of CNP-based capacitor. This improvement in energy density can be attributed to the enhancement of effective cation diffusion in the SPANI−CNP composites.

A SPANI−CNP-based supercapacitor with AN/SAN ratio of 1 was employed to test for the cycle stability though galvanostatic charge and discharge in current density of 2 A·g^{-1} for 5000 cycles. Figure 11 shows 96.1% retention in capacitance after 5000 cycles, indicating excellent cycle stability in this supercapacitor.

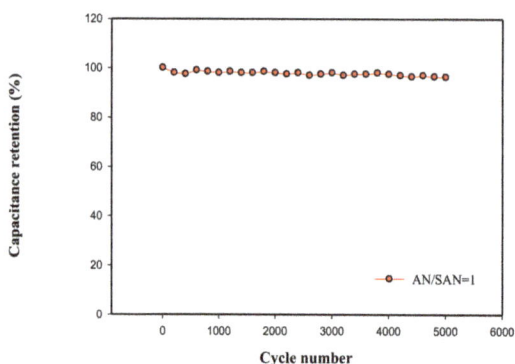

Figure 11. Cycling stability of SPANI−CNP-based supercapacitor at current density of 2 A·g^{-1}.

4. Conclusions

We electrodeposited self-doped polyaniline (SPANI) onto carbon nanoparticle (CNP) film to form a SPANI−CNP composite by electropolymerization and improved the capacitive performance of the CNP electrode through increasing its conductivity and psuedocapacitive characteristic. SPANI with AN/SAN ratio of 1 exhibited the best performance among these AN/SAN ratios. In galvanostatic charge and discharge, the SPANI−CNP composite with AN/SAN ratio of 1 exhibited maximum specific capacitance of 273.3 F·g^{-1} compared to 133.3 F·g^{-1} in the CNP electrode, the specific energy density and power density of SPANI−CNP composite with AN/SAN ratio of 1 were 7.3 Wh·Kg^{-1} and 5996.8 W·Kg^{-1} compared to 3.3 Wh·Kg^{-1} and 2997 W·Kg^{-1} for the CNP electrode. Moreover, the capacitance retention was 96.3% after 5000 cycles in galvanostatic charge and discharge. Our results have shown that the SAN can replace other external doping and have the same supercapacitive performance as polyaniline/carbon composites. The SPANI−CNP composite with AN/SAN ratio of 1 had a superior supercapacitive performance due to its Faradic contribution and its efficient charge-transfer and ion transport between electrolyte and electrode. These results demonstrated that the SPANI−CNP composites have practical potential for electrochemically stable supercapacitors.

Acknowledgments: Authors greatly acknowledged financial support from the Ministry of Science and Technology in Taiwan (MOST 106-2221-E-390-025 and MOST 106-2622-E-390-001-CC2).

Author Contributions: Po-Hsin Wang performed the experiments; Chien-Hsin Yang, Wen-Churng Lin, and Tzong-Liu Wang conceived and designed the experiments; Po-Hsin Wang, Hung-Yin Lin, Mei-Hwa Lee, and Chien-Hsin Yang analyzed the data; Hung-Yin Lin, Chien-Hsin Yang, and Tzong-Liu Wang contributed reagents/materials/analysis tools; Chien-Hsin Yang, Po-Hsin Wang, Wen-Churng Lin, Hung-Yin Lin, Mei-Hwa Lee, and Tzong-Liu Wang wrote the paper.

Conflicts of Interest: The authors declare no conflict of interest. The founding sponsors had no role in the design of the study; in the collection, analyses, or interpretation of data; in the writing of the manuscript, and in the decision to publish the results.

References

1. Burke, A. Ultracapacitors: Why, how, and where is the technology. *J. Power Sources* **2000**, *91*, 37–50. [CrossRef]
2. Arico, A.S.; Bruce, P.; Scrosati, P.B.; Tarascon, J.M.; Van Schalkwijk, W. Nanostructured materials for advanced energy conversion and storage devices. *Nat. Mater.* **2005**, *4*, 366–377. [CrossRef] [PubMed]
3. Soudan, P.; Gaudet, J.; Guay, D.; Bélanger, D.; Schulz, R. Electrochemical properties of ruthenium-based nanocrystalline materials as electrodes for supercapacitors. *Chem. Mater.* **2002**, *14*, 1210–1215. [CrossRef]
4. Choi, B.Y.; Chung, I.J.; Chun, J.H.; Ko, J.M. Electrochemical characteristics of dodecylbenzene sulfonic acid-doped polyaniline in aqueous solutions. *Synth. Met.* **1999**, *99*, 253–256. [CrossRef]
5. Izadi-Najafabadi, A.; Yasuda, S.; Kobashi, K.; Yamada, T.; Futaba, D.N.; Hatori, H.; Yumura, M.; Iijima, S.; Hata, K. Extracting the full potential of single-walled carbon nanotubes as durable supercapacitor electrodes operable at 4 V with high power and energy density. *Adv. Mater.* **2010**, *22*, E235–E241. [CrossRef] [PubMed]
6. Nishihara, H.; Simura, T.; Kobayashi, S.; Nomura, K.; Berenguer, R.; Ito, M.; Uchimura, M.; Iden, H.; Arihara, K.; Ohma, A.; et al. Oxidation-resistant and elastic mesoporous carbon with single-layer graphene walls. *Adv. Funct. Mater.* **2016**, *26*, 6418–6427. [CrossRef]
7. Frackowiak, E.; Khomenko, V.; Jurewicz, K.; Lota, K.; Beguin, F. Supercapacitors based on conducting polymers/nanotubes composites. *J. Power Sources* **2006**, *153*, 413–418. [CrossRef]
8. Liu, W.; Yan, X.; Xue, Q. Multilayer hybrid films consisting of alternating graphene and titanium dioxide for high performance supercapacitors. *J. Mater. Chem. C* **2013**, *1*, 1413–1422. [CrossRef]
9. Peng, H.; Ma, G.; Sun, K.; Zhang, Z.; Yang, Q.; Rana, F.; Lei, Z. A facile and rapid preparation of highly crumpled nitrogen-doped graphene-like nanosheets for high-performance supercapacitors. *J. Mater. Chem. A* **2015**, *3*, 13210–13214. [CrossRef]
10. Wang, J.; Gao, Z.; Li, Z.; Wang, B.; Yan, Y.; Liu, Q. Green synthesis of graphene nanosheets/ZnO composites and electrochemical properties. *J. Solid State Chem.* **2011**, *184*, 1421–1427. [CrossRef]
11. Ryu, K.S.; Lee, Y.; Han, K.S.; Parka, Y.J.; Kang, M.G.; Park, N.G.; Chang, S.H. Electrochemical supercapacitor based on polyaniline doped with lithium salt and active carbon electrodes. *Solid State Ion.* **2004**, *175*, 765–768. [CrossRef]
12. Yan, J.; Liu, J.; Fan, Z.; Wei, T.; Zhang, L. High-performance supercapacitor electrodes based on highly corrugated graphene sheets. *Carbon* **2012**, *50*, 2179–2188. [CrossRef]
13. Ramadoss, A.; Ki, S. Improved activity of a grapheme–TiO$_2$ hybrid electrode in an electrochemical supercapacitor. *Carbon* **2013**, *63*, 434–445. [CrossRef]
14. Grover, S.; Goel, S.; Sahu, V.; Singh, G.; Sharma, R.K. Asymmetric supercapacitive characteristics of PANI embedded holey graphene nanoribbons. *ACS Sustain. Chem. Eng.* **2015**, *3*, 1460–1469. [CrossRef]
15. Winokur, M.J.; Mattes, B.R. Polyaniline as viewed from a structural perspective. *J. Reinf. Plast. Compos.* **1999**, *18*, 875–884. [CrossRef]
16. Yang, C.H.; Chih, Y.K.; Cheng, H.E.; Chen, C.H. Nanofibers of self-doped polyaniline. *Polymer* **2005**, *46*, 10688–10698. [CrossRef]
17. Wei, Y.; Focke, W.W.; Wnek, G.E.; Ray, A.; MacDiarmid, A.G. Synthesis and electrochemistry of alkyl ring-substituted polyanilines. *J. Phys. Chem.* **1989**, *93*, 495–499. [CrossRef]
18. Yang, C.H.; Wen, T.C. Polyaniline derivative with external and internal doping via electrochemical copolymerization of aniline and 2,5-diaminobenzenesulfonic acid on IrO$_2$-coated titanium electrode. *J. Electrochem. Soc.* **1994**, *141*, 2624–2632. [CrossRef]

nanomaterials

MDPI

Article

Electrospun Composites of Polycaprolactone and Porous Silicon Nanoparticles for the Tunable Delivery of Small Therapeutic Molecules

Steven J. P. McInnes [1], Thomas J. Macdonald [2], Ivan P. Parkin [2], Thomas Nann [3] and Nicolas H. Voelcker [4,5,*]

[1]   Future Industries Institute, University of South Australia, Mawson Lakes 5095, Australia;
      steven.mcinnes@unisa.edu.au
[2]   Department of Chemistry, University College London, London WC1E 6BT, UK;
      tom.macdonald@ucl.ac.uk (T.J.M.); i.p.parkin@ucl.ac.uk (I.P.P.)
[3]   MacDiarmid Institute for Advanced Materials and Nanotechnology, School of Chemical and Physical
      Sciences, Victoria University of Wellington, Wellington 6012, New Zealand; thomas.nann@vuw.ac.nz
[4]   Monash Institute of Pharmaceutical Sciences, Monash University, Parkville 3052, Australia
[5]   Commonwealth Scientific and Industrial Research Organisation (CSIRO), Clayton 3168, Australia
*   Correspondence: nicolas.voelcker@monash.edu.au; Tel.: +61-3-9902-9097

Received: 28 February 2018; Accepted: 27 March 2018; Published: 29 March 2018

Abstract: This report describes the use of an electrospun composite of poly(ε-caprolactone) (PCL) fibers and porous silicon (pSi) nanoparticles (NPs) as an effective system for the tunable delivery of camptothecin (CPT), a small therapeutic molecule. Both materials are biodegradable, abundant, low-cost, and most importantly, have no known cytotoxic effects. The composites were treated with and without sodium hydroxide (NaOH) to investigate the wettability of the porous network for drug release and cell viability measurements. CPT release and subsequent cell viability was also investigated. We observed that the cell death rate was not only affected by the addition of our CPT carrier, pSi, but also by increasing the rate of dissolution via treatment with NaOH. This is the first example of loading pSi NPs as a therapeutics nanocarrier into electronspun PCL fibers and this system opens up new possibilities for the delivery of molecular therapeutics.

Keywords: porous silicon; drug delivery; electrospinning; poly(ε-caprolactone)

1. Introduction

Synthetic or natural biocompatible polymers are commonly considered as candidates to develop scaffolds for tissue engineering [1]. Effective scaffolds for tissue engineering need to consist of materials that are highly porous, fibrous, biocompatible, biodegradable, and cause no harm to the immune system. It is also important that the complex function of the scaffold mimics the extracellular matrix (ECM), the vital model in providing structural and biochemical support to human cells.

Using compatible biomaterials such as poly(ε-caprolactone) (PCL), chitosan (CS), or gelatin (GEL) is a common approach to engineer a variety of tissue types. The biodegradable polyester PCL is among the most studied scaffolds in tissue engineering and is approved by the United States Food and Drug Administration (FDA, Silver Spring, MD, USA) [2]. PCL has high plasticity, ductility, and ester linkages, allowing for a slow degradation rate from hydrolysis, which are all useful characterization tools for these scaffolds [3]. Biocompatible nanofibers (NFs) produced by electrospinning have been a popular scaffold material due to their similar characteristics to the ECM. PCL NFs have been shown to form suitable interwoven porous scaffolds [4–7], which assist in the connection of tissues and vessels. These NFs have also been shown to be appropriate structures to mimic the ECM, due to their ability to promote the adhesion and proliferation of cells [8,9]. Additionally, PCL has previously been shown to support a wide range of range of cell types, and its biodegradable features render it an excellent candidate for carrying therapeutic molecules [5].

Porous silicon (pSi) is another example of a biomaterial, and has several unique properties making it attractive for both in vivo and ex vivo applications [10]. pSi is formed by the anodization of crystalline silicon wafers in hydrofluoric acid (HF). pSi nanostructures are easily tailored by altering the wafer resistivity, HF concentration, and current density [11]. Furthermore, pore sizes may be tuned in diameter from a few nanometers to several microns, achieving surface areas of up to 800 m^2/g [12]. The biocompatibility of pSi has been demonstrated in several animal models [13]. Bimbo et al. showed that the biodistribution of pSi nanoparticles (pSi NPs) does not induce toxicity or inflammatory responses while displaying excellent in vivo stability in rats [14]. Ex vivo applications for pSi include implantable biosensors [15], cancer diagnostics [16], and wound healing [17]. Previously investigated in vitro functionalized pSi NPs and immune cell interactions revealed that even at concentrations of 250 mg/mL, there were no significant cytotoxicity effects [18]. The surface chemistry of pSi is important to load different drugs where a range of in vivo and in vitro studies have demonstrated the biocompatibility of surface-modified pSi [19]. We have reported on composite pSi and PCL membranes implanted into the subconjunctival space of rats [20]. These membranes did not erode or cause inflammatory responses in the tissue surrounding the implant and there was also no evidence of vascularization. Our more recent work investigated the utility of surface-modified pSi membranes as a scaffold for the transfer of oral mucosal cells to the eye. We found that pSi scaffolds supported and retained transplanted rat oral mucosal epithelial cells both in vitro and in vivo [21]. Furthermore, pSi is not limited to biological applications whereby our previous work has also shown its adaptability in solar energy conversion. This verifies pSi as a versatile nanomaterial which can be used in a variety of applications [22].

Electrospinning uses an electrical charge to draw fine fibers from a liquid polymer solution to a charged collector plate. These fibers are typically one-dimensional (1-D) porous structures, which range from submicron to several nanometers in size [23]. Electrospun fibers and in particular NFs offer promising properties such as large surface area to volume ratio, surface flexibility, and superior tensile strength [24,25]. NFs are exciting candidates for a range of applications such as drug delivery [26], biosensing [27], photovoltaic devices [28], and energy storage [29]. An imperative requirement for drug delivery systems is the generation of porous scaffolds to accommodate cells in guiding their growth and regeneration in three dimensions (3-D). This can be achieved using multiple layers of electrospun materials, which produce 3-D fibrillar NF mats [30]. NFs that are used in drug delivery typically follow one of two designs: firstly, for NFs with homogenous structures, the drug or target molecule may be dispersed throughout the fibrous polymer matrix. Secondly, core-shell or coaxial NFs may be fabricated whereby a polymer covers the matrix carrying the drug [31,32]. The drug diffusion mechanism and drug release kinetics are different for the two designs. For example, in homogenous NFs, the drug must travel progressively long distances in order to diffuse throughout the outer edge of the NFs, meaning that the rate of release typically decreases with time. In contrast, core-shell NF systems enable stable diffusion rates of the drugs due to the structure of the system; however, their preparation is much more complex. A core advantage of using NFs is their structure (diameter, density, and thickness), which can be easily tailored by varying the process parameters or combining other nanomaterials within the fibrous structures [33,34]. Furthermore, fiber diameter is tunable and dependent on polymer molecular weight, sol-gel concentration, flow rate, applied voltage, needle tip size, and hydrolytic degradation of NFs [35].

Herein, we combine the favorable properties of both pSi and PCL to manufacture drug eluting composite electrospun PCL fibers containing pSi NPs. We investigate the release kinetics of the small molecular drug camptothecin (CPT). In an effort to control the release kinetics, we investigate the difference of release properties of materials loaded directly with CPT and those with the equivalent CPT amount preloaded into pSi NPs. We also investigate the effect of pretreating the PCL fibers with sodium hydroxide (NaOH) to enhance their wettability and degradation rate [36].

2. Materials and Methods

2.1. Chemicals

Hydrofluoric acid (HF) 48% (Merck), dichloromethane (CH_2Cl_2, Labserv, analytical grade, 99.5%), and polycarolactone (80,000 average m wt) were purchased from Sigma-Aldrich and used as received. Methanol (Merck, analytical grade, 99.5%), DCM (Chemsupply, analytical grade, 99.5%), and ethanol (Ajax, absolute, 100%) were used without further purification. *N,N*-dimethylformamide (DMF, EMD Chemicals, Overijse, Belgium) was purified via standard laboratory protocols including drying over $MgSO_4$ followed by distillation at reduced pressure [36]. Milli-Q water was obtained from an Advantage A10 water purification system provided by Merck Millipore (water resistivity of 18.2 MΩcm at 25 °C, Total Organic Carbon (TOC) < 5 ppb). Dulbecco's phosphate buffered saline (PBS) solution and fluorescein isothiocyanate (97.5%, FITC) were purchased from Sigma Aldrich and used as received.

2.2. Fabrication of pSi NPs

pSi NPs were fabricated from p-type Si wafers (boron-doped, resistivity 0.0008–0.0012 Ω cm, <100>) supplied by Siltronix (Archamps, France). The wafer was anodized in an 18-cm^2 etching cell in 3:1 HF:ethanol (v/v) solution with a square wave form comprising an initial current density of 50 mA/cm^2 for 7.3 s and a second current density of 400 50 mA/cm^2 for 0.4 s. This two-step cycle was repeated continuously for 1 h, generating a pSi film with alternating low and high porosity layers. The etched layer was removed from the Si substrate via electropolishing in 1:20 HF:EtOH at 4 mA/cm^2 for 4 min and 10 s. Subsequently, the pSi membrane was sonicated for 16 h in DMSO to generate chemically oxidized pSi NPs. These nanoparticles were sized by passing through a 220-nm polytetrafluoroethylene (PTFE) syringe filter, followed by the collection of the pellet after centrifugation at 22,000× *g*. This filtration and centrifugation allowed for the removal of large and small nanoparticles and facilitated the harvest of reasonably uniformly sized NPs that permanently remained in solution.

2.3. CPT Loading of pSi NPs and PCL Composite Materials

pSi NPs were placed into an Eppendorf tube, to which a known volume (50–200 µL) of CPT solution of approximately 4.6 mg/mL CPT in dry distilled DMF was added. This mixture was allowed to incubate for 2 h before the particles were dried under vacuum (10 mm Hg) in a desiccator. The total loadings were calculated based on the mass of pSi placed into the tube originally; for example, a typical loading used 200 µL of 4.6 mg/mL CPT and 15.0 mg of pSi, resulting in a loading of 61.3 µg of CPT per mg of pSi. Each individual batch loading was used to convert the release amounts into percentages.

To load the PCL NFs, a predetermined mass of pSi was added to the spinning solution based on the loading of the pSi NPs. The final loading of CPT in the spinning solution was normalized to contain 0.63 mg of CPT for both the PCL + CPT materials and the PCL + pSi-CPT materials. In the above example, 10.3 mg of CPT loaded pSi NPs would be added to the spinning solution. To ensure a homogeneous distribution of the pSi NPs in the spinning solution, the particles were briefly sonicated in a small amount of acetone before injection into the spinning solution. After injection, the solutions were stirred as best as possible and then placed into the syringe for spinning.

2.4. Water Contact Angle (WCA) Measurements

The WCA was measured by placing a 1-µL drop of water on the sample surface and capturing a digital image using a Panasonic Super Dynamic wv-BP550 Closed Circuit TV camera. The contact angle measurements were analyzed by Scion Image for Windows Framegrabber software (Beta version 4.0.2).

2.5. Electrospinning of PCL and NaOH Treatments

In a typical synthesis, a 5-mL solution of 10% PCL in acetone with pSi NPs was electrospun from a 23 G stainless steel needle. The mass of pSi added to the electrospinning solution was dependent on the loading of the particular batch of pSi; however, for all of this work the mass of pSi used was balanced to facilitate a loading equivalent to 0.63 mg of CPT in the pSi + CPT composite material. Hence, every material spun and tested for drug release contained the same initial loading of CPT. The needle was then connected to a high-voltage supply (Gamma High Voltage Research, Ormand Beach, FL, USA). The solution was fed at a rate of 0.5 mL/h using a syringe pump (PHD 2000, Harvard Apparatus, Holliston, MA, USA). A piece of flat aluminum foil was placed 10 cm below the tip of the needle as a collector plate. The voltage for the electrospinning was set to 15 kV and was conducted in a controlled temperature environment (25 °C).

2.6. Scanning Electron Microscopy (SEM)

SEM was performed on an FEI Quanta 450 FEG environmental SEM fitted with a Secondary Electron Detector (SED) detector and operated at 30 keV with a spot number of 2. To help facilitate the dissipation of charge build-up, samples were coated with a 5-nm thick layer of Pt prior to analysis, according to our standard laboratory protocol.

2.7. X-ray Photoelectron Spectroscopy (XPS)

XPS measurements were recorded on a Thermo Scientific K-alpha spectrometer with monochromatic Al-K$_\alpha$ radiation at University College London. This involved obtaining a monatomic depth profile of the NFs using an ion beam to etch layers of the surface revealing subsurface information. The etching was performed for 200 s, which was calibrated to penetrate ~50 nm into the surface. Peak positions were calibrated to carbon (285 eV) and plotted using the CasaXPS and qtiplot software.

2.8. Energy-Dispersive X-ray Spectroscopy (EDX)

EDX was obtained using an Oxford Instruments UTW Energy Dispersive Spectroscopy (EDS) detector running ISIS software. The EDS detector was ran through a JEOL JSM-6301F Field Emission SEM.

2.9. Fluorescence Microscopy

Fluorescence microscopy was performed on an Eclipse 50*i* microscope equipped with a D-FL universal epi-fluorescence attachment and a 100-W mercury lamp (Nikon Instruments, Tokyo, Japan). Fluorescence images were captured with a Charged-Coupled Device (CCD) camera (Nikon Instruments, Tokyo Japan), using the following fluorescence filters. Blue channel (violet excitation, blue emission): excitation: 385–400 nm (bandpass, 393 CWL), dichromatic mirror: 435–470 nm (bandpass), and barrier filter wavelength: 450–465 nm (bandpass, 458 CWL). Green channel (blue excitation, green emission): excitation: 475–490 nm (bandpass, 483 CWL), dichromatic mirror: 500–540 nm (bandpass), and barrier filter wavelength: 505–535 nm (bandpass, 520 CWL). Red channel (green excitation, orange/red emission): excitation: 545–565 nm (bandpass, 555 CWL), dichromatic mirror: 570–645 nm (bandpass), and barrier filter wavelength: 580–620 nm (bandpass, 600 CWL). Images were analyzed using NIS-elements v3.07 software (Nikon Instruments, Tokyo, Japan).

2.10. Drug Release

CPT release was monitored via fluorimetry and performed on a Perkin Elmer Instruments LS55 luminescence spectrometer with an excitation wavelength of 340 nm and emission wavelength of 434 nm. The slit width was set to 3 nm and the photomultiplier voltage was set to 775 V. The cumulative release data of CPT into 3 mL of PBS was monitored over a 17-h period. Release rates were calculated from the slope of release curves obtained. The actual amount of CPT released was calculated with reference to a calibration curve and normalized to the surface area of the sample to give the amount of CPT released per cm^2. This allowed the CPT release data to be directly compared between each of the samples. A minimum of three release curves was averaged to produce the release curves. The release of CPT was performed in PBS at pH 7.4 and pH 1.8. Despite the low solubility of CPT in aqueous solutions (14.2 ± 2.9 μM) [37], sink conditions were maintained for all release experiments (maximum release concentrations were below 1.4 μM).

2.11. Cytotoxicity Assay

SH-SY5Y cells (human neuroblastoma cells), were cultured in Dulbecco's Modified Eagle's Medium (DMEM) supplemented with 10% FBS, 100 U/mL penicillin, and 100 μg/mL streptomycin (Invitrogen) as previously described. SH-SY5Y cells were placed in wells of a 96-well plate at 15,000 cells per well. After one day, the cultured cells were incubated with the prepared CPT-loaded and CPT-free PCL NFs (the following set of samples was used: PCl only, PCl + CPT, PCl + pSi-CPT). PCl NFs in DMEM (10% FBS, 100 U/mL penicillin, and 100 μg/mL streptomycin) were incubated for 24, 48, and 96 h at 37 °C and 5% CO_2 prior to cell viability measurements. Controls were generated by incubating SH-SY5Y cells in DMEM without a PCl NF for an identical incubation period.

To determine the effect of the microparticles treatment on cell viability, the percentage of live and dead cells, lactate dehydrogenase (LDH) released in culture supernatants was measured using an established assay (Abcam LDH-Cytotoxicity Assay Kit II) according to the manufacturer's instructions. After 24, 48, and 96 h of incubation with microparticles, 100 μL of the cell suspension was removed and centrifuged at 600× *g* for 10 min, and the supernatant was transferred to a 96-well plate. To each well, 100 μL LDH reaction mix (Abchem) was added. After 30 min of incubation at room temperature, the absorbance at 450 nm was measured. All experiments were repeated at least three times.

3. Results and Discussion

3.1. Material Characterization

The characterization of pSi NPs was performed in our earlier work [38]. Briefly, the particle size was found to be in the range of 161 ± 58 nm and the particles possessed a pore size of 33 ± 7 nm. The characterization of the pSi NPs can be found in the Supporting Information (ESI, Figure S1).

Electrospun materials were removed from the Al foil backing and folded until they were eight layers thick. Subsequently, 3-mm diameter NF discs were punched from the PCL sheet using a hole punch. The NFs were of homogeneous size and thickness and on average weighed 1.43 ± 0.66 mg. NaOH is known to speed up the hydrolysis of the PCL polymer backbone [36]. We calculated an average % mass loss of 3.5% PCL per hour of exposure to NaOH. We anticipated the base treatment to enhance the speed at which the PCL degraded and hence control the release rate of the CPT. Scanning electron microscopy (SEM) of the NFs is shown below (Figures 1 and 2) and indicates that the PCL NF diameter ranges between several hundred nanometers and a few microns, which is consistent with previous reports [39,40]. However, from the SEM characterization in Figure 1, it is clear that the use of NaOH on the prepared fibrous networks destroyed the smaller fibers (<500 nm) in the network, leaving behind larger fibers which in fact slowed the release of CPT rather than speeding it up.

X-ray photoelectron spectroscopy (XPS) of the PCL and PCL with pSi (PCL + pSi) NFs was performed to determine the presence of pSi in the PCL fibers; the resulting XPS of the Si 2p spectra is shown below in Figure 3. Given the small amounts of silicon expected to be preset, and its likelihood of appearing on the XPS spectra as a surface contaminant, XPS for PCL NFs and pSi PCL NFs was obtained by measuring a depth profile of each of the samples. XPS depth profiling involved using an ion beam to etch off subsequent layers, revealing subsurface information. The depth profile corresponded to a penetration of roughly 50 nm for each NF sample. The more intense Si 2p signal in the pSi PCL NFs suggests that the pSi NPs are well embedded into the surface of the NFs. Figure 3 shows characteristic Si 2p peaks at 103 eV for both pSi PCL and PCL NF samples. This binding energy corresponds to oxidized pSi ($pSiO_2$), which is expected after the pSi NPs are sonicated in DMSO and subject to oxidation (see experimental section). The small Si 2p signal present in the PCL NF sample is a likely a contaminant, since siloxane compounds are widely used as lubricants and release agents in polymer materials. Siloxanes have a strong tendency to migrate towards polymer surfaces and at small bulk concentrations, resulting in segregation to the outer surface and causing high surface contamination [41]. The contaminant is typically characterized as a broad, low intense siloxane peak, which is consistent with the high resolution Si 2p spectrum shown in Figure 3 [41]. Furthermore, given that the depth profile measurements only showed an increase in Si 2p signal for the pSi PCL NF sample, this suggests that the pSi NPs are well within the pSi PCL NF network. The binding energies with their chemical assignments are shown in the Electronic Supporting Information (ESI), Table S1. EDX analysis before and after NaOH treatment also corroborated the XPS findings (ESI, Figure S2).

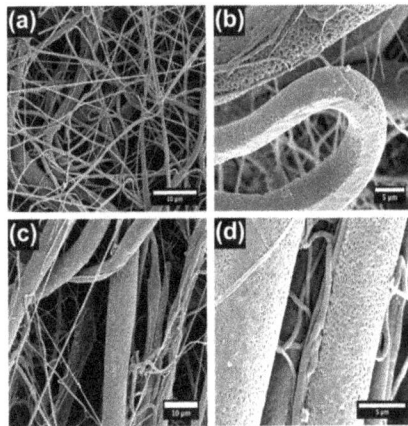

Figure 1. SEM characterization of poly(ε-caprolactone) nanofibers (PCL NFs) with and without treatment with NaOH. (**a**,**b**) No NaOH treatment and (**c**,**d**) 24-h NaOH treatment.

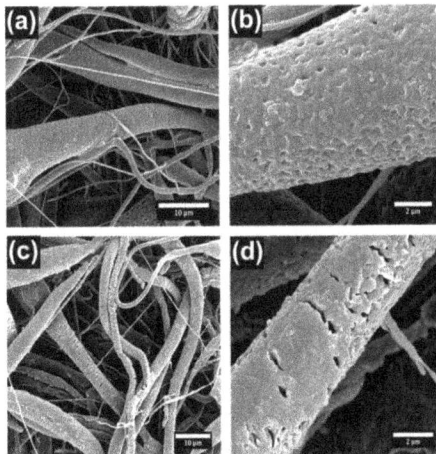

Figure 2. SEM characterization of PCL + porous silicon (pSi) NFs with and without treatment with NaOH. (**a**,**b**) No NaOH treatment; (**c**,**d**) 24-h NaOH treatment.

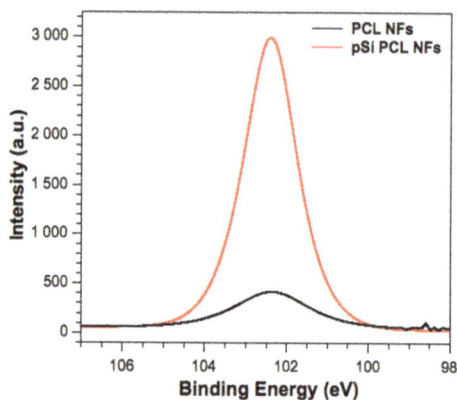

Figure 3. High-resolution Si 2p X-ray photoelectron spectroscopy (XPS) characterization of PCL and PCL + pSi NFs.

The static water contact angle (WCA) of the PCL fibers was 113.5 ± 8.4° (Figure 4a). However, after a 1-h treatment of the PCL with NaOH, the PCL NFs completely absorbed the 1-μL droplet placed onto the surface (Figure 3b). This complete wetting of the water droplet into the PCL NFs could be due to both the increase of the hydrophilicity of the fibers [36] and the opening up of the porous mesh due to the breakage of smaller NFs and the pitting of the thicker NFs after the NaOH treatment (Figure 4).

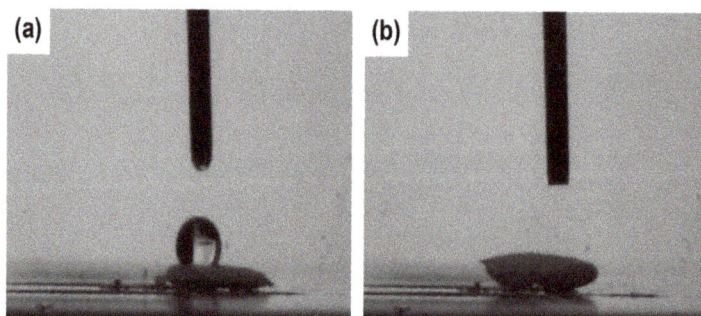

Figure 4. Water contact angle (WCA) measurements of (**a**) PCL without NaOH treatment and (**b**) PCL with 1-h NaOH treatment.

3.2. Drug Loading and Release

Throughout this work, the mass of pSi added to the electrospinning solution was balanced to facilitate the loading of CPT, according to our methodology as detailed in the experimental section. Every material spun and tested for drug release contained the same initial loading of CPT.

Fluorescence microscopy of the loaded and unloaded materials (Figure 5) below was conducted in an attempt to track and visualize the location of the pSi NPs and CPT throughout the PCL + pSi composite materials. Figure 5 demonstrates the fluorescence of the pSi NPs, the PCL only, and the two different composite preparations in the DAPI (CPT-sensitive) channel. It is clear that the pSi NPs and the PCL themselves showed no auto-fluorescence. However, upon the addition of CPT, there was a notable increase in the fluorescence in the DAPI channel for both CPT only and pSi-CPT containing samples, indicating that CPT was located throughout the network of both sample preparations.

CPT drug release was monitored from the PCL + CPT and PCL + pSi-CPT composite materials both with and without the NaOH treatment (Figures 6a and 7a). The PCL only control material did not show any release or interference in the CPT signal measured at 434 nm. Before NaOH treatment, the PCL + pSi-CPT showed significantly higher % release from the pSi-containing composite compared to the PCL + CPT (Figure 6a), possibly due to the pSi creating weak points in the surface of the PCL fibers, which are more susceptible to hydrolysis and aid in speeding the dissolution of the PCL fiber.

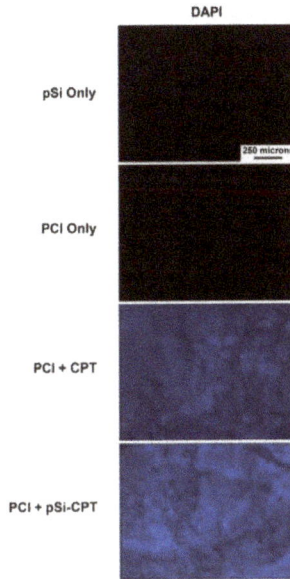

Figure 5. Fluorescence microscopy images of the original pSi NPs, PCL materials, and the two composite materials in the DAPI channels. All images are taken at the same exposure settings to facilitate qualitative comparison.

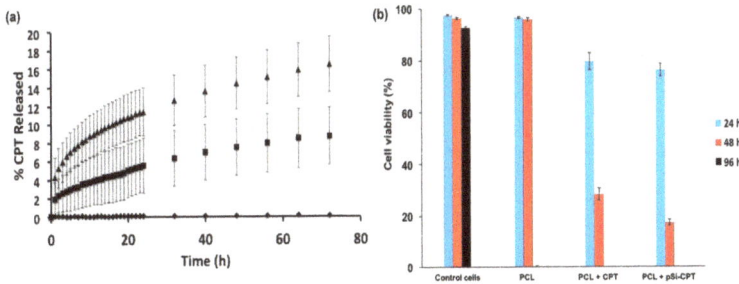

Figure 6. (a) Release curves of camptothecin (CPT) from PCL + CPT and PCL + pSi-CPT composite materials without NaOH treatment. ♦ = PCL control material, ■ = PCl + CPT, ▲ = PCl + pSi-CPT; (b) Viability of SH-SY5Y human neuroblastoma cells upon incubation with PCL and PCL + pSi-composite materials with or without CPT loading ($n \geq 3$).

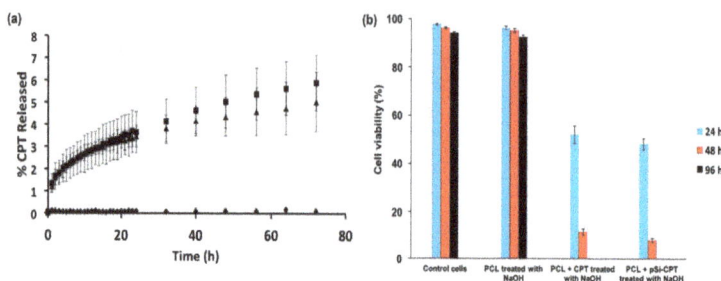

Figure 7. (a) Release curves of CPT from (a) PCL + CPT composites and (b) PCL + pSi-CPT composite materials after 1-h NaOH treatment. ◆ = PCL control material, ■ = PCL + CPT, ▲ = PCL + pSi-CPT. Viability of SH-SY5Y human neuroblastoma cells upon incubation with PCL and PCL + pSi-composite materials with or without CPT loading and NaOH treatment ($n \geq 3$).

The untreated PCL + CPT composite and the 1-h NaOH treated PCL + CPT composite showed release rates of 8.72 ± 3.13% and 5.90 ± 1.25%, respectively (Figures 6a and 7a). The PCL only materials (Figures 6a and 7a) did not display any release in either the treated or untreated case. The same materials also possessed average release rates of 0.121%/h and 0.082%/h, respectively. The untreated PCL + pSi-CPT composite and the 1-h treated PCL + pSi-CPT composite materials showed final release percentages of 16.48 ± 3.04% and 5.03 ± 1.35%, respectively (Figures 6a and 7a). The same materials also possessed average release rates of 0.229%/h and 0.070%/h, respectively.

Modeling of the release data (Table 1) revealed that all of the materials best fit the Higuchi model. This indicates that the release followed a mostly diffusion-controlled process out of the PCL fiber networks. Using the Ritger-Peppas model to assess the release mechanism via the 'n' exponent revealed that the 'n' exponent was always between 0.45 and 1. Therefore, we can assume that the release mechanism is a non-Fickian (anomalous) diffusion for all samples. This indicates a mixture of release from the CPT diffusing out of the fibers and some additional effects from the PCL polymer molecules as the water infiltrates the polymer network.

Table 1. Modeling of the release kinetics and mechanism.

Material	Zero-Order	First Order	Higuchi	Hixson-Crowell	Ritger-Peppas	'n'
PCL + CPT	0.899	0.925	0.992	0.558	0.981	0.74
PCL + CPT 1-h NaOH	0.890	0.909	0.991	0.559	0.987	0.70
PCL + pSi-CPT	0.806	0.886	0.966	0.425	0.999	0.63
PCL + pSi-CPT 1-h NaOH	0.823	0.843	0.974	0.531	0.998	0.67

3.3. Cell Viability Assay

Figures 6b and 7b show the cell viability in an LDH assay for all PCL composite samples over a 4-day time course. The control (untreated) neuroblastoma cells and the cells treated with PCL only and PCL only with NaOH treatment (1 h) showed virtually 100% cell viability over the 4-day time period (Figures 6b and 7b). This confirms that neither the pSi nor any residual NaOH left behind after treatment caused any cytotoxic effect (at least for the LDH assay). In stark contrast, the CPT-loaded PCL and PCL + pSi materials (Figure 6b) showed a rapid onset of decreased viability within the first day and this only increased over the full 4-day time course. The viability of the cells exposed to PCL + CPT and PCL + pSi-CPT samples not pretreated with NaOH showed decreased viability from 80% to 28% and 76% to 17%, respectively, from day 1 to 2 (Figure 6b). Both samples had 0% cell viability at day 4. Similarly, the same two sample sets pretreated for 1 h in NaOH dropped in viability from 52% to 12% and 48% to 8%, respectively, from day 1 to 2 (Figure 7b). Again, both samples had 0% cell viability at day 4 (Figure 7b). These biological results demonstrate that the cell death rate was not necessarily affected by the addition of the pSi as a carrier for the CPT, but it was affected by increasing the dissolution rate via treatment with NaOH.

3.4. General Discussion

PCL materials were successfully fabricated via electrospinning and were able to encase the model drug CPT or pSi NPs preloaded with CPT. To the best of our knowledge, this is the first time pSi NPs have been spun through PCL fibers. This simple entrapment of small molecular drugs in a carrier such as pSi could lead to the development of electrospun constituents that are able to contain sensitive payloads that would normally not be able to be incorporated directly into these materials. NFs with small diameters have a large surface area per unit mass, and this work is just one example of how particles and biological structures can be isolated and protected inside NFs, while remaining accessible for use when needed. Herein, we show that NFs loaded with pSi can be used as convenient carriers for the tunable delivery of small therapeutic molecules [42].

The release rates of the small molecular model drug, CPT, can also be modulated both via incorporation into pSi particles or by direct and intentional degradation (incubation with NaOH) of the PCL scaffold. The ability to affect the release rate of CPT from these materials also gives rise to a significant difference in the cell cytotoxicity of the released CPT on SH-SY5Y cells. This can be exploited for the tunable delivery of therapeutic payloads to areas such as non-resectable tumors. Furthermore, this work opens up the possibility of delivering a range of therapeutic molecules from electrospun fibers, potentially leading to the development of smart scaffolds that can allow for cell growth and infiltration whilst facilitating the release of cell growth-stimulating biological factors.

4. Conclusions

PCL polymeric fibers could be combined both directly with a small molecular drug payload and a particulate carrier system for the drug payload. Tunable drug release and subsequent cell death could be achieved by tuning the composition of the PCL composites or pretreating them with NaOH to generate a porosity that enhances dissolution. Our methodology may open the possibility of protecting small therapeutic molecules within a porous carrier, such as porous silicon, during the high voltage spinning process for later sustained long-term release applications.

Supplementary Materials: The following are available online at http://www.mdpi.com/2079-4991/8/4/205/s1, Figure S1: Representative pSi NP TEM microscopy images. Particles were found to be typically 161 ± 58 nm with a pore size of 33 ± 7 nm. Table S1. XPS of PCL and PCL + pSi composite discs. Figure S2: (Top) EDX spectra for PCl fibers after NaOH treatment. (Bottom) EDX spectra for PCL fibers containing pSi NPs after NaOH treatment.

Acknowledgments: T.J.M. would like to acknowledge the Ramsay Memorial Fellowship Trust for their financial support. I.P.P. would like to acknowledge the EPSRC for financial support (EP/M015157/1).

Author Contributions: S.J.P.M. performed the pSi NP fabrication, loading, and polymer preparations. S.J.P.M. also performed the pressing of the PCl into discs and the WCA, SEM, and drug release analysis. T.J.M. performed the electrospinning experiments as well as the EDX and XPS analysis. N.H.V. and T.N. conceived and designed the experiments.

Conflicts of Interest: The authors of this paper declare no conflicts of interest.

References

1. Endres, M.; Hutmacher, D.; Salgado, A.; Kaps, C.; Ringe, J.; Reis, R.; Sittinger, M.; Brandwood, A.; Scantz, J.T. Osteogenic Induction of Human Bone Marrow-Derived Mesenchymal Progenitor Cells in Novel Synthetic Polymer—Hydrogel Matrices. *Tissue Eng.* **2003**, *9*, 689–701. [CrossRef] [PubMed]
2. Ulery, B.D.; Nair, L.S.; Laurencin, C.T. Biomedical Applications of Biodegradable Polymers. *J. Polym. Sci. Part B Polym. Phys.* **2011**, *49*, 832–864. [CrossRef] [PubMed]
3. Gomes, S.R.; Rodrigues, G.; Martins, G.G.; Roberto, M.A.; Mafra, M.; Henriques, C.M.R.; Silva, J.C. In vitro and in vivo evaluation of electrospun nanofibers of PCL, chitosan and gelatin: A comparative study. *Mater. Sci. Eng. C* **2015**, *46*, 348–358. [CrossRef] [PubMed]

4. Chakrapani, V.Y.G.A.; Giridev, V.R.; Madhusoothanan, M.; Sekaran, G. Electrospinning of type I collagen and PCL nanofibers using acetic acid. *J. Appl. Polym. Sci.* **2012**, *125*, 3221–3227. [CrossRef]

5. Cho, S.J.; Jung, S.M.; Kang, M.; Shin, H.S.; Youk, J.H. Preparation of hydrophilic PCL nanofiber scaffolds via electrospinning of PCL/PVP-b-PCL block copolymers for enhanced cell biocompatibility. *Polymer* **2015**, *69*, 95–102. [CrossRef]

6. Prosecká, E.; Rampichová, M.; Litvinec, A.; Tonar, Z.; Králíková, M.; Vojtová, L.; Kochová, P.; Plencner, M.; Buzgo, M.; Míková, A.; et al. Collagen/hydroxyapatite scaffold enriched with polycaprolactone nanofibers, thrombocyte-rich solution and mesenchymal stem cells promotes regeneration in large bone defect in vivo. *J. Biomed. Mater. Res. A* **2015**, *103*, 671–682. [CrossRef] [PubMed]

7. Sridhar, R.; Madhaiyan, K.; Sundarrajan, S.; Gora, A.; Venugopal, J.R.; Ramakrishna, S. Cross-linking of protein scaffolds for therapeutic applications: PCL nanofibers delivering riboflavin for protein cross-linking. *J. Mater. Chem. B* **2014**, *2*, 1626–1633. [CrossRef]

8. Khil, M.-S.; Bhattarai, S.R.; Kim, H.-Y.; Kim, S.-Z.; Lee, K.-H. Novel fabricated matrix via electrospinning for tissue engineering. *J. Biomed. Mater. Res. B Appl. Biomater.* **2005**, *72*, 117–124. [CrossRef] [PubMed]

9. Yoshimoto, H.; Shin, Y.M.; Terai, H.; Vacanti, J.P. A biodegradable nanofiber scaffold by electrospinning and its potential for bone tissue engineering. *Biomaterials* **2003**, *24*, 2077–2082. [CrossRef]

10. Woodhead Publishing Series in Biomaterials. In *Porous Silicon for Biomedical Applications*; Woodhead Publishing: Cambridge, UK, 2014.

11. McInnes, S.J.P.; Szili, E.J.; Al-Bataineh, S.A.; Xu, J.; Alf, M.E.; Gleason, K.K.; Short, R.D.; Voelcker, N.H. Combination of iCVD and Porous Silicon for the Development of a Controlled Drug Delivery System. *ACS Appl. Mater. Interfaces* **2012**, *4*, 3566–3574. [CrossRef] [PubMed]

12. Canham, L. *Handbook of Porous Silicon*; Springer International Publishing: Basel, Switzerland, 2014.

13. Low, S.P.; Voelcker, N.H.; Canham, L.T.; Williams, K.A. The biocompatibility of porous silicon in tissues of the eye. *Biomaterials* **2009**, *30*, 2873–2880. [CrossRef] [PubMed]

14. Bimbo, L.M.; Sarparanta, M.; Santos, H.A.; Airaksinen, A.J.; Mäkilä, E.; Laaksonen, T.; Peltonen, L.; Lehto, V.-P.; Hirvonen, J.; Salonen, J. Biocompatibility of Thermally Hydrocarbonized Porous Silicon Nanoparticles and their Biodistribution in Rats. *ACS Nano* **2010**, *4*, 3023–3032. [CrossRef] [PubMed]

15. Acquaroli, L.N.; Kuchel, T.; Voelcker, N.H. Towards implantable porous silicon biosensors. *RSC Adv.* **2014**, *4*, 34768–34773. [CrossRef]

16. Sailor, M.J.; Park, J.-H. Hybrid Nanoparticles for Detection and Treatment of Cancer. *Adv. Mater.* **2012**, *24*, 3779–3802. [CrossRef] [PubMed]

17. Pace, S.; Vasani, R.B.; Cunin, F.; Voelcker, N.H. Study of the optical properties of a thermoresponsive polymer grafted onto porous silicon scaffolds. *New J. Chem.* **2013**, *37*, 228–235. [CrossRef]

18. Santos, H.A.; Mäkilä, E.; Airaksinen, A.J.; Bimbo, L.M.; Hirvonen, J. Porous silicon nanoparticles for nanomedicine: Preparation and biomedical applications. *Nanomedicine* **2014**, *9*, 535–554. [CrossRef] [PubMed]

19. Salonen, J.; Laitinen, L.; Kaukonen, A.M.; Tuura, J.; Björkqvist, M.; Heikkilä, T.; Vähä-Heikkilä, K.; Hirvonen, J.; Lehto, V.-P. Mesoporous silicon microparticles for oral drug delivery: Loading and release of five model drugs. *J. Control. Release* **2005**, *108*, 362–374. [CrossRef] [PubMed]

20. Kashanian, S.; Harding, F.; Irani, Y.; Klebe, S.; Marshall, K.; Loni, A.; Canham, L.; Fan, D.; Williams, K.A.; Voelcker, N.H.; et al. Evaluation of mesoporous silicon/polycaprolactone composites as ophthalmic implants. *Acta Biomater.* **2010**, *6*, 3566–3572. [CrossRef] [PubMed]

21. Irani, Y.D.K.S.; McInnes, S.J.P.; Jasieniak, M.; Voelcker, N.H.; Williams, K.A. Oral Mucosal Epithelial Cells Grown on Porous Silicon Membrane for Transfer to the Rat Eye. *Sci. Rep.* **2017**, *7*, 10042. [CrossRef] [PubMed]

22. Chandrasekaran, S.; McInnes, S.J.P.; Macdonald, T.J.; Nann, T.; Voelcker, N.H. Porous silicon nanoparticles as a nanophotocathode for photoelectrochemical water splitting. *RSC Adv.* **2015**, *5*, 85978. [CrossRef]

23. Macdonald, T.J.; Xu, J.; Elmas, S.; Mange, Y.J.; Skinner, W.M.; Xu, H.; Nann, T. NiO Nanofibers as a Candidate for a Nanophotocathode. *Nanomaterials* **2014**, *4*, 256–266. [CrossRef] [PubMed]

24. Batmunkh, M.; Macdonald, T.J.; Shearer, C.J.; Baterdene, M.; Wang, Y.; Biggs, M.J.; Parkin, I.P.; Nann, T.; Shapter, J.G. Carbon Nanotubes in TiO$_2$ Nanofiber Photoelectrodes for High Performance Perovskite Solar Cells. *Adv. Sci.* **2017**, *4*, 1600504. [CrossRef] [PubMed]

25. Huang, Z.-M.; Zhang, Y.-Z.; Kotaki, M.; Ramakrishna, S. A review on polymer nanofibers by electrospinning and their applications in nanocomposites. *Compos. Sci. Technol.* **2003**, *63*, 2223–2253. [CrossRef]

26. Kim, Y.-J.; Ebara, M.; Aoyagi, T. A Smart Hyperthermia Nanofiber with Switchable Drug Release for Inducing Cancer Apoptosis. *Adv. Funct. Mater.* **2013**, *23*, 5753–5761. [CrossRef]
27. Squires, A.H.; Hersey, J.S.; Grinstaff, M.W.; Meller, A. A Nanopore-Nanofiber Mesh Biosensor to Control DNA Translocation. *J. Am. Chem. Soc.* **2013**, *135*, 16304–16307. [CrossRef] [PubMed]
28. Macdonald, T.J.; Tune, D.D.; Dewi, M.R.; Gibson, C.T.; Shapter, J.G.; Nann, T. A TiO$_2$ Nanofiber–Carbon Nanotube-Composite Photoanode for Improved Efficiency in Dye-Sensitized Solar Cells. *ChemSusChem* **2015**. [CrossRef]
29. Ambroz, F.; Macdonald, T.J.; Nann, T. Trends in Aluminium Based Intercalation Batteries. *Adv. Energy Mater.* **2017**, *7*, 7. [CrossRef]
30. Delalat, B.; Harding, F.; Gundsambuu, B.; De-Juan-Pardo, E.M.; Wunner, F.M.; Wille, M.-L.; Jasieniak, M.; Malatesta, K.A.L.; Griesser, H.J.; Simula, A.; et al. 3D printed lattices as an activation and expansion platform for T cell therapy. *Biomaterials* **2017**, *140*, 58–68. [CrossRef] [PubMed]
31. Jiang, H.; Wang, L.; Zhu, K. Coaxial electrospinning for encapsulation and controlled release of fragile water-soluble bioactive agents. *Drug Deliv. Res. Asia Pac. Reg.* **2014**, *193*, 296–303. [CrossRef] [PubMed]
32. Xu, X.; Chen, X.; Ma, P.; Wang, X.; Jing, X. The release behavior of doxorubicin hydrochloride from medicated fibers prepared by emulsion-electrospinning. *Eur. J. Pharm. Biopharm.* **2008**, *70*, 165–170. [CrossRef] [PubMed]
33. Hrib, J.; Sirc, J.; Hobzova, R.; Hampejsova, Z.; Bosakova, Z.; Munzarova, M.; Michalek, J. Nanofibers for drug delivery—Incorporation and release of model molecules, influence of molecular weight and polymer structure. *Beilstein J. Nanotechnol.* **2015**, *6*, 1939–1945. [CrossRef] [PubMed]
34. Saraf, A.; Baggett, L.S.; Raphael, R.M.; Kasper, F.K.; Mikos, A.G. Regulated non-viral gene delivery from coaxial electrospun fiber mesh scaffolds. *J. Control. Release* **2010**, *143*, 95–103. [CrossRef] [PubMed]
35. Dias, J.C.; Ribeiro, C.; Sencadas, V.; Botelho, G.; Ribelles, J.L.G.; Lanceros-Mendez, S. Influence of fiber diameter and crystallinity on the stability of electrospun poly(L-lactic acid) membranes to hydrolytic degradation. *Polym. Test.* **2012**, *31*, 770–776. [CrossRef]
36. Araujo, J.V.; Martins, A.; Leonor, I.B.; Pinho, E.D.; Reis, R.L.; Neves, N.M. Surface controlled biomimetic coating of polycaprolactone nanofiber meshes to be used as bone extracellular matrix analogues. *J. Biomater. Sci. Polym. Ed.* **2008**, *19*, 1261–1278. [CrossRef] [PubMed]
37. Sætern, A.M.; Nguyen, N.B.; Bauer-Brandl, A.; Brandl, M. Effect of hydroxypropyl-a-cyclodextrin-complexation and pH on solubility of camptothecin. *Int. J. Pharm.* **2004**, *284*, 61–68. [CrossRef] [PubMed]
38. Stead, S.O.; McInnes, S.J.P.; Kireta, S.; Rose, P.D.; Jesudason, S.; Rojas-Canales, D.; Warther, D.; Cunin, F.; Durand, J.-O.; Drogemuller, C.J.; et al. Manipulating human dendritic cell phenotype and function with targeted porous silicon nanoparticles. *Biomaterials* **2018**, *155*, 92–102. [CrossRef] [PubMed]
39. Baker, S.R.; Banerjee, S.; Bonin, K.; Guthold, M. Determining the mechanical properties of electrospun poly-e-caprolactone (PCL) nanofibers using AFM and a novel fiber anchoring technique. *Mater. Sci. Eng. C* **2016**, *59*, 203–212. [CrossRef] [PubMed]
40. Chew, S.Y.; Hufnagel, T.C.; Lim, C.T.; Leong, K.W. Mechanical properties of single electrospun drug-encapsulated nanofibres. *Nanotechnology* **2006**, *17*, 3880–3891. [CrossRef] [PubMed]
41. Oran, U.; Ünveren, E.; Wirth, T.; Unger, W.E. Poly-dimethyl-siloxane (PDMS) contamination of polystyrene (PS) oligomers samples: A comparison of time-of-flight static secondary ion mass spectrometry (TOF-SSIMS) and X-ray photoelectron spectroscopy (XPS) results. *Appl. Surf. Sci.* **2004**, *227*, 318–324. [CrossRef]
42. Reneker, D.H.; Yarin, A.L. Electrospinning jets and polymer nanofibers. *Polymer* **2008**, *49*, 2387–2425. [CrossRef]

nanomaterials

MDPI

Article

Solvothermal Synthesis of a Hollow Micro-Sphere LiFePO₄/C Composite with a Porous Interior Structure as a Cathode Material for Lithium Ion Batteries

Yang Liu [1], Jieyu Zhang [1,2], Ying Li [1,*], Yemin Hu [1,*], Wenxian Li [1], Mingyuan Zhu [1], Pengfei Hu [1], Shulei Chou [3] and Guoxiu Wang [4]

[1] Laboratory for Microstructures, School of Materials Science and Engineering, Shanghai University, Shanghai 200072, China; yangliu8651@shu.edu.cn (Y.L.); zjy6162@staff.shu.edu.cn (J.Z.); shuliwx@shu.edu.cn (W.L.); zmy@shu.edu.cn (M.Z.); hpf-hqx@shu.edu.cn (P.H.)

[2] Shanghai Key Laboratory of Modern Metallurgy & Materials Processing, School of Materials Science and Engineering, Shanghai University, Shanghai 200072, China

[3] Institute for Superconducting and Electronic Materials, University of Wollongong, Wollongong, NSW 2522, Australia; shulei@uow.edu.au

[4] Department of Chemistry and Forensic Science, University of Technology Sydney, Sydney, NSW 2007, Australia; Guoxiu.Wang@uts.edu.au

* Correspondence: liying62@shu.edu.cn (Y.L.); huyemin@shu.edu.cn (Y.H.); Tel.: +86-021-5633-8874 (Y.L.)

Received: 26 September 2017; Accepted: 24 October 2017; Published: 3 November 2017

Abstract: To overcome the low lithium ion diffusion and slow electron transfer, a hollow micro sphere LiFePO₄/C cathode material with a porous interior structure was synthesized via a solvothermal method by using ethylene glycol (EG) as the solvent medium and cetyltrimethylammonium bromide (CTAB) as the surfactant. In this strategy, the EG solvent inhibits the growth of the crystals and the CTAB surfactant boots the self-assembly of the primary nanoparticles to form hollow spheres. The resultant carbon-coat LiFePO₄/C hollow micro-spheres have a ~300 nm thick shell/wall consisting of aggregated nanoparticles and a porous interior. When used as materials for lithium-ion batteries, the hollow micro spherical LiFePO₄/C composite exhibits superior discharge capacity (163 mAh g⁻¹ at 0.1 C), good high-rate discharge capacity (118 mAh g⁻¹ at 10 C), and fine cycling stability (99.2% after 200 cycles at 0.1 C). The good electrochemical performances are attributed to a high rate of ionic/electronic conduction and the high structural stability arising from the nanosized primary particles and the micro-sized hollow spherical structure.

Keywords: lithium ion battery; lithium iron phosphate; solvothermal method; micro hollow sphere

1. Introduction

As one of the most promising polyanion-type cathode materials for high-power Li-ion batteries in electric vehicles (EVs) and energy storage, the olivine-structured LiFePO₄ has been extensively studied all over the world since 1997. LiFePO₄ possess numerous appealing features, such as low cost, environmentally benign, good thermal stability, and perfect flat voltage profile at 3.45 V vs. Li⁺/Li [1–3]. However LiFePO₄ suffers from two main disadvantages: low ionic-electronic conductivity (10^{-9}–10^{-10} S cm⁻¹) and limited lithium ion diffusion channel (one-dimensional path along the b-axis), which significantly restricts the rate performance when attempting fast charging or discharging [4,5]. Tremendous efforts have been exerted to overcome the electronic and ionic transport restriction by optimizing morphology [6–8], reducing particle size [9–13], decorating the surface with electricallyconducting agents [14–16], and doping the host framework with supervalent cations [17–21]. Among these strategies, size reduction and carbon coating are considered as effective methods to improve the performance of LiFePO$_4$, but there still remain some fundamental and technical challenges. Firstly, such electrodes can suffer from undesirable reactions between electrode and electrolyte, which may arise from the high surface area and high surface energies of materials with nano size [9,22]. On the other hand, it is difficult to obtain a full carbon layer coating the surface of LiFePO$_4$ particles, so electrons may not reach all the positions where Li$^+$ ion intercalation takes place during the charge/discharge process. This can lead to unwanted polarization [14,23–25].

Recent research indicates that carbon-coated LiFePO$_4$ cathode materials made in the form of hollow microspheres composed of nanoparticles are able to solve the above problems [26–30]. Nanosized particles of LiFePO$_4$ can decrease the Li$^+$ ion migration lengths and improve the kinetics of LiFePO$_4$, while the hollow micro-spherical structure prevents collapse in the long-term cycles and improves the contact between electrode and electrolyte. The most prominent improvement caused by the structure is the small potential drop (or rise) observed just after discharge (or charge) begins at various charge rates. The hydrothermal or solvothermal methods have been demonstrated to be one of most popular routes to fabricate hollow micro-spherical LiFePO$_4$ composites with many advantages: mild synthesis conditions, high purity, highly degree of crystallinity, narrow particle size distribution, and vast size ranges. Huang et al. prepared hollow micro-spherical structure LiFePO$_4$ with a superior discharge capacity of 101 mAh g^{-1} at 20 C by using carbon spheres as hard templates via the hydrothermal method [31]. However, the hard template methods are time-consuming and costly because of the need for initial synthesis and the final removal of the template. A hollow microspherical LiFePO$_4$ material synthesized via hydrothermal method with CTAB as surfactant was reported Lee et al. [32], in which the hollow LiFePO$_4$/C composite exhibited a superior energy density of 312 W h g^{-1} at 10 C. However, many impurities were taken into the resultant composite and the balance between morphologies of particles and impurities were studied in another article [33]. Yang et al. [29] also reported monodisperse LiFePO$_4$ hollow micro-spheres as high performance cathode materials via solvothermal synthesis using spherical Li$_3$PO$_4$ as the self-sacrificed template, PEG 600 as a surfactant, and FeCl$_2$·4H$_2$O as the Fe^{2+} source in an EG (ethylene glycol) medium. All the above spherical LiFePO$_4$ particles were hollow, which results in a waste of the large interior volume, bringing a loss of tap density.

In this work, using inexpensive raw material NH$_4$H$_2$PO$_4$ as P source and CTAB as surfactant in an ethylene glycol (EG) medium, a facile solvothermal route was adopted to synthesize hollow structured LiFePO$_4$ spheres with a porous interior which resembles the yolk inside an egg but is composed of carbon-coat primary nanoparticles. The porous spherical shell can permit electrolyte to permeate quickly invade into the interior of the hollow sphere to obtain good contact with the porous "yolk" region, leading to an increasing of specific area, which would facilitate both electronic and lithium ionic diffusion. The as-obtained hollow LiFePO$_4$/C microsphere, which was designated as LFP-A, shows a high reversible specific capacity of 163 mAh g^{-1} at 0.1 C, as well as excellent rate capability and cycling performance.

2. Results and Discussion

2.1. Microstructure

The crystal structures of the two composites were investigated by X-ray diffraction. As shown in Figure 1, both samples display the orthorhombic phase with a space group Pmnb (62) (JCPDS card No. 81-1173). The profiles of the reflection peaks are quite narrow, indicating the high crystallinity of the samples. No obvious diffraction peaks of impurity phases (e.g., Fe$_2$O$_3$, Li$_3$PO$_4$, and Fe$_2$P) are observed in the patterns, which indicate the high purity of the samples.

The scanning electron microscopy (SEM) images of the two samples are shown in Figure 2. Uniform spherical particles with a diameter of approximately 1~3 μm are observed in Figure 2a, and the insert picture shows a broken sphere, exhibiting the likely typical hollow spherical structure of the LFP-A sample. A magnified image of an incomplete microsphere is shown in Figure 2b, indicating that the spherical particles of LFP-A sample is not solid but hollow, with an interior core similar to an egg yolk. It was also observed that the ~300 nm thick porous wall and yolk-interior of the hollow spheres are both composed of ~100 nm primary particles and the yolk is not solid, but porous. The transmission electron microscopy (TEM) micrograph shown in Figure 2c displays a general image of the hollow microspheres of LFP-A sample with a diameter of 1.2 μm. The high-resolution transmission electron microscope (HRTEM) image of the primary particles shown in Figure 2d demonstrates that the nanosized particles are well-crystallized and conformably coated by about 2.5 nm of carbon layers. In contrast, the particles of the sample (LFP-B) synthesized without CTAB display a nanosized spindle-like morphology, which can be clearly observed in the SEM (Figure 2e) and TEM images (Figure 2f). Owing to the novel structure, the LFP-A sample obtains a tap density of 1.2 g cm^{-3}, significantly higher than the 0.9 g cm^{-3} measured for the nanosized LFP-B sample. The tap density of the LFP-A sample is similar to the value of other microspherical LiFePO$_4$ composites with a solid interior reported recently [34–36].

The sizes of the primary particles of LFP-A and the whole particles of LFP-B synthesized via our solvothermal route are smaller than that of the particles synthesized by the hydrothermal route in our previous work [21]. The size reduction of particles is attributed to the organic solvent EG which has two hydroxyl groups in its molecule, capable of weak linking to the LFP nanocrystallites via hydrogen bonds, thus inhibiting the growth of the crystals [15]. Comparing the micrographs of the two samples, it can be seen that the morphologies of the powders are completely changed with CTAB involved in the process. The anticipated mechanism for the formation of LiFePO$_4$/C particles is shown in Figure 3. An EG-water mixture with a certain fraction of CTAB will self-assemble to form CTAB rod-like micelles in an ordered arrangement. With the decrease of the weight fraction of CTAB as the mixture was introduced into the Li$_3$PO$_4$ suspension, the growth of CTAB micelles evolves from a hexagonal crystalline phase to an isotropic solution phase [37,38]. Then, hexagonal crystalline micelles will serve as templates in the formation of Fe^{2+}-Li$_3$PO$_4$ composite units which will nucleate further LiFePO$_4$ material and the composite units grow in size to form nano-particles. As the reaction proceeds, the particle groups assemble to from spheres owing to the residual CTAB molecules, which gather particles around themselves. After calcination, hollow LiFePO$_4$ spheres can be obtained because of the decomposition of CTAB. Therefore, the structure changes, from spindle-like solid particles to hollow microspheres, can be attributed to the presence of the CTAB surfactant. The CTAB in the solvothermal synthesis also works as a carbon source, and is converted to conductive amorphous carbon after heat treatment. Meanwhile, it acts as a carbonaceous-reducing agent during heating to prevent oxidation of Fe^{2+} to Fe^{3+}.

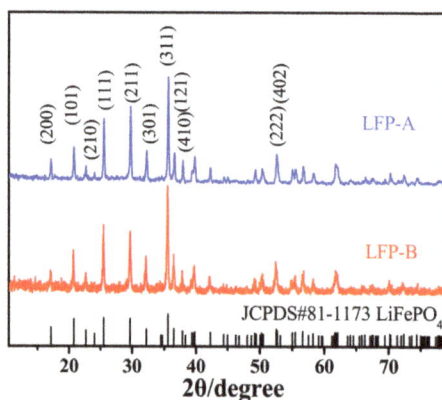

Figure 1. XRD patterns of LFP-A and LFP-B.

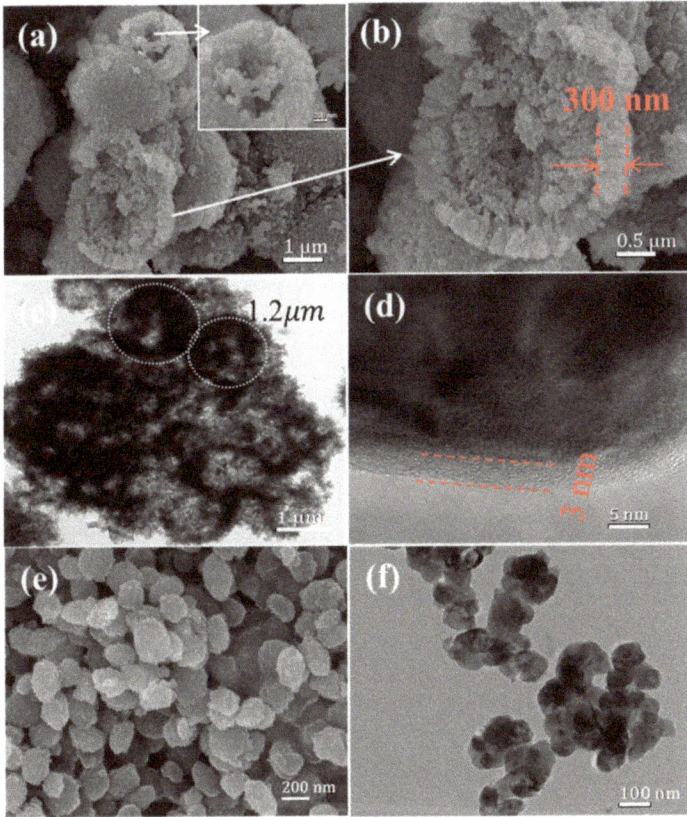

Figure 2. (**a,b**) SEM images of LFP-A; (**c,d**) TEM images of LFP-A; and (**e**) SEM and (**f**) TEM images of LFP-B.

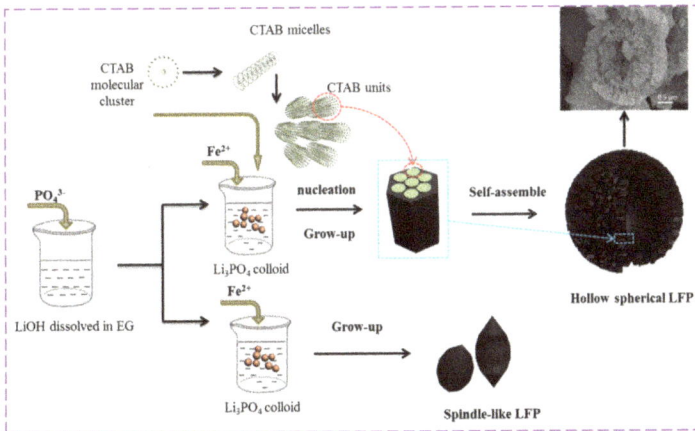

Figure 3. Schematic illustration of the formation of LiFePO$_4$/C particles.

Figure 4a shows the nitrogen adsorption-desorption isotherms curves for LFP-A sample. The typical-IV curve and H3 hysteresis indicate the presence of mesopores in the as-prepared LiFePO$_4$ sample. According to the curve, the specific Brunauer-Emmett-Teller (BET) surface area can be calculated to 31.27 m^2 g^{-1}, and the high specific surface area could provide more sites for lithium ion insertion/extraction, resulting in good electrochemical performance.

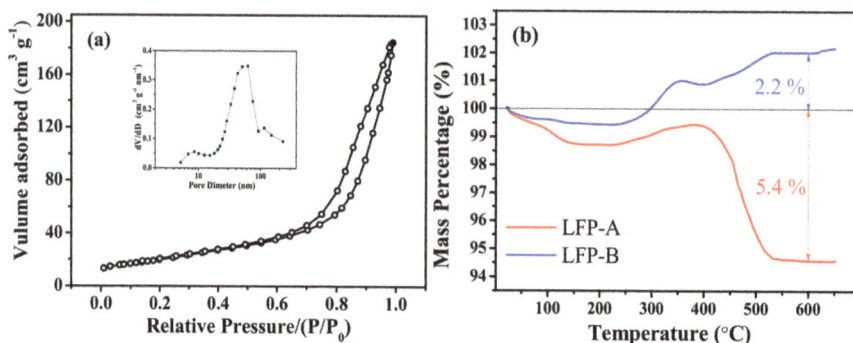

Figure 4. (a) Nitrogen adsorption-desorption isotherm curve of LFP-A sample; and (b) TG curves of the samples with a heating rate of 10 °C min^{-1} in air.

The carbon content in the two samples was estimated by thermogravimetric (TG) measurement in air and the curves are shown in Figure 4b. It should be mentioned that, in the temperature range of 250–500 °C, the olivine LiFePO$_4$ can be oxidized to Li$_3$Fe$_2$(PO$_4$)$_3$ and Fe$_2$O$_3$, corresponding to a theoretical weight gain of 5.07%. While the carbon in the samples starts to be oxided to CO$_2$ gas above 350 °C, leading to a weight loss, and is burnt out completely above 500 °C, based on the following Equation (1) [39]:

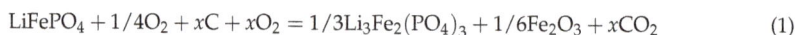

$$\text{LiFePO}_4 + 1/4\text{O}_2 + x\text{C} + x\text{O}_2 = 1/3\text{Li}_3\text{Fe}_2(\text{PO}_4)_3 + 1/6\text{Fe}_2\text{O}_3 + x\text{CO}_2 \tag{1}$$

where x denotes the carbon content in the composite. By noting the small deviation in mass upon heating, the percentages of carbon in LFP-A and LFP-B are calculated to be 10.77% and 2.87%, respectively. The high carbon content of LFP-A is attributed to the addition of CTAB surfactant and the formation of protective hollow sphere shapes with a porous interior structure.

2.2. Electrochemical Performances

The samples were tested as a cathodic material for lithium ion rechargeable batteries. The loading of electrode material is ca. 1.73 mg cm^{-2}, with a 8:1:1 weight ratio of LiFePO$_4$/C, acetylene carbon black and poly(vinylidene fluoride). The electrode area is 1.67 cm^2 with a density of around 1.2 g cm^{-3}. The charge/discharge curves of the synthesized LFP samples are shown in Figure 5. LFP-A composite delivers a high specific capacity of 163 mAh g^{-1} at 0.1 C-rate (29.5 × 10^{-3} mA cm^{-2}) with a long and flat plateau, which is 95.9% of the theoretical capacity (170 mAh g^{-1}). While the LFP-B sample delivers a relatively lower capacity of 153 mAh g^{-1}. The LFP-A shows a smaller voltage difference of ca. 31 mV between the charge and discharge plateaus which means it has smaller polarization, i.e., the interfacial electrostatic behavior of the hollow microspherical LiFePO$_4$/C composite excel that of the nanosized spindle-like LiFePO$_4$/C, which exhibits a voltage difference of ca. 52 mV.

The cycling performance of the two samples measured between 2.0 and 4.2 V at a current density of 0.1 C-rate (29.5 × 10^{-3} mA cm^{-2}) are displayed in Figure 6. It can be seen that LFP-A and LFP-B samples both exhibit good capacity retention with continuous charge-discharge processes. The capacity of LFP-B sample declines from 152 mAh g^{-1} at the initial cycle to 145 mAh g^{-1} at the 200th cycle, with a capacity loss of 4.5%, while the capacity decay of LFP-A sample after 200 cycles is only 0.8%. The sample composed of hollow microspheres exhibits a better cycling performance, which can be attributed to the microporous hollow spherical structure increasing the contact area between the active materials and the electrolyte.

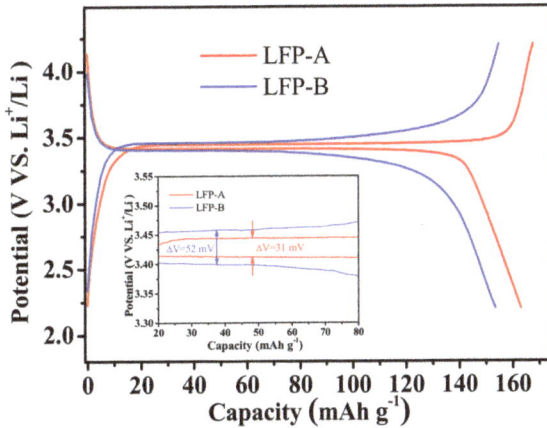

Figure 5. Charge/discharge profiles of the two samples at 0.1 C rate. The inset shows the magnified flat region.

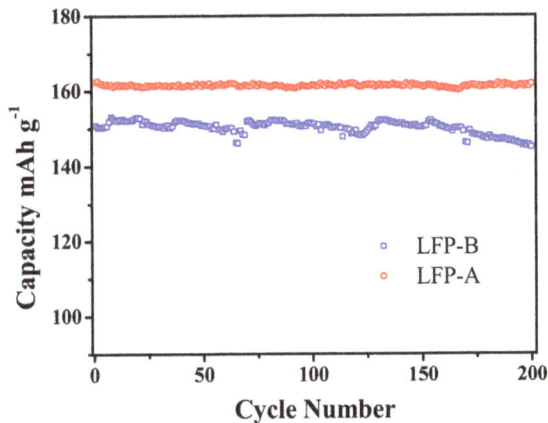

Figure 6. Discharge capacities during continuous cycling of samples at 0.1 C.

Charge and discharge tests at various current rates ranging from 0.1 C to 10 C were performed and the results are shown in Figure 7. The LFP-A sample exhibits a superior rate performance, and capacities as high as 163, 152, 145, 137, and even 118 mA h g^{-1} could be obtained at 0.1, 0.5, 1, 5, and 10 C rates, respectively. The reversible capacity can be recovered and maintained 161.9 mA h g^{-1} when the current rate was again returned to 0.1 C. The corresponding values of the sample synthesized without CTAB are 154, 142, 125, 117, and 102 mA h g^{-1}, respectively. The microspherical secondary architecture of hollow LiFePO$_4$/C is, hence, suggested to be more favorable for the diffusion of both electrons and ions.

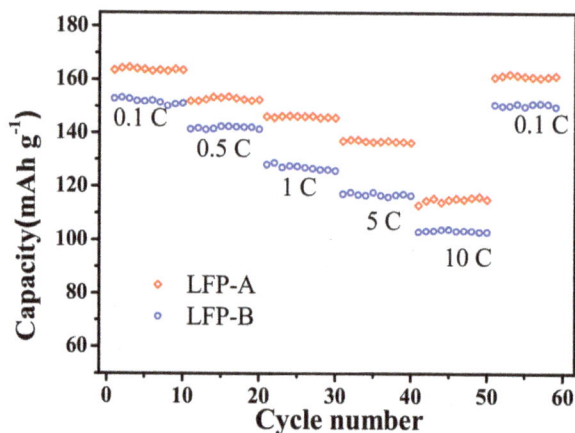

Figure 7. Cyclability of discharge capacities at various charge rates.

To obtain more information on the electrochemical properties of the two samples, cyclic voltammetry (CV) tests were carried out at a scanning rate of 0.1 mV s^{-1} in the potential range of 2.2–4.5 V. As shown in Figure 8, the LFP-A sample showed a more symmetric shape and sharper peak profiles. It exhibits a pair of anodic and cathodic peaks at 3.52 V and 3.34 V, respectively, resulting in a potential shift of 0.18 V. In the case of the LFP-B composite synthesized without CTAB, the anodic and cathodic peaks were found at 3.54 V and 3.33 V, respectively, with a potential shift of 0.21 V. These peaks correspond to the extraction and insertion of lithium ions and the smaller potential shift illustrates a weaker polarization and possibly an easier kinetic process for LFP-A. It can also be found that the current flow at redox peaks of LFP-A is higher than that in the LFP-B sample. The anodic/cathodic current peaks of LFP-A were about 1.1 mA, while the value for the other sample was about 0.2 mA. These peaks correspond to the extraction and insertion of lithium ions. According to the Randles-Sevcik equation [40], it can be concluded that the Li$^+$ ion diffusion in the hollow spherical LiFePO$_4$/C is faster than that in the spindle-like LiFePO$_4$/C.

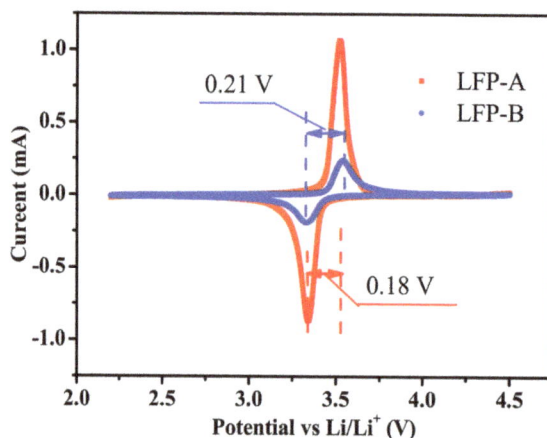

Figure 8. Cyclic voltammograms of the two composites at a scan rate of 0.1 mV s^{-1}.

The electrochemical impendance spectra (EIS) plots of the samples are shown in Figure 9a. Both materials exhibit a semicircle in the high-frequency region and a straight line in the low-frequency region. The numerical value of the diameter of the semicircle on the Z_{re} axis is approximately equal to the charge transfer resistance (*Rct*) [41]. The straight line is attributed to the diffusion of the lithium ions into the bulk of the electrode material, or so-called Warburg diffusion. The lithium-ion diffusion coefficient could be calculated by using the following Equation (2) [42]:

$$D = R^2T^2/2A^2n^4F^4c^2\sigma^2 \qquad (2)$$

where R is the gas constant (8.314 J mol^{-1} K^{-1}), T is the absolute temperature, A is the surface area of the cathode, n is the number of electrons per molecule during oxidization, F is the Faraday constant (96,500 C mol^{-1}), c is the concentration of lithium ion (7.69 × 10^{-3} mol cm^{-3}), and σ is the Warburg factor, which is related with Z_{re} via Equation (3):

$$Z_{re} = R_s + R_{ct} + \sigma\omega^{-1/2} \qquad (3)$$

where Z_{re} represents the real part of the resistance in the low frequency region and ω is the corresponding frequency. The relationships between Z' and the square root of frequency in the low-frequency region of the two samples are shown in Figure 9b. From the two curves, Li$^+$ ion diffusion coefficients (D) of sample LFP-A and LFP-B electrodes are calculated to be 2.32 × 10^{-12} and 5.05 × 10^{-13} cm^2 s^{-1}, respectively. The result indicates that Li ion diffusion in LFP-A is nearly five times faster than that in LFP-B. The values of the diffusion coefficients for the two samples are higher than the literature value for pristine LiFePO$_4$ (~10^{-14} cm^2 s^{-1}) [17]. This further confirms that the hollow micro-sphere structure is helpful to the rapid Li ion transport.

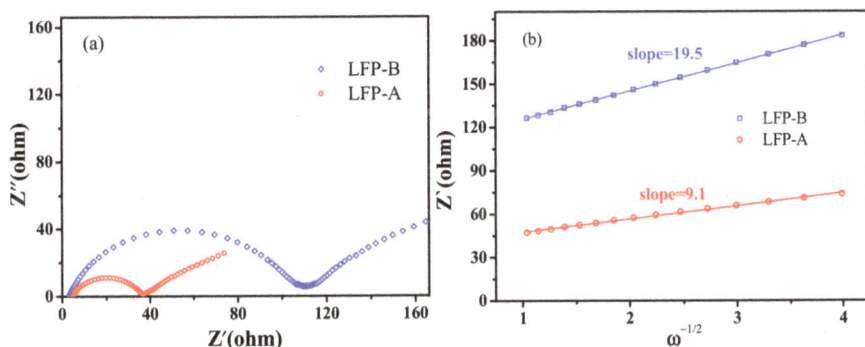

Figure 9. (a) Electrochemical impedance spectroscopy (EIS) of the two composites and (b) the relationships between Z' and $\omega^{-1/2}$ in low-frequency regions.

Comparing the test results of the two samples, it clearly demonstrates the more excellent electrochemical performance of the LiFePO$_4$/C composite with hollow micro-sphere shells and porous interior structure. The distinct improvement in electrochemical performance can be attributed to their successfully designed structural features: (1) the hollow micro-spheres with porous interior structure to prevent structural collapse in long-term cycling by supplying enough space for the change of volume in the extraction and insertion of Li$^+$ ions; (2) the nanoparticles comprising the microspheres to promote good electron transfer and enhance the accessibility of lithium ions [9]; and (3) the porous hollow microspherical structure also provides good contact with electrolyte which can flood the interior of the hollow spheres [43]. Even LFP-B, synthesized in the absence of CTAB surfactant, also displays an excellent electrochemical performance, better than that of pristine LiFePO$_4$ synthesized via the hydrothermal route in our previous work, and the electrical performance presented in this work is compared with other known hollow spherical LiFePO$_4$ composites reported in the literatures, as shown in Table 1. Its high-scoring properties can be attributed to the nano-dimension particle size and the uniform carbon coating layer.

Table 1. Comparison of hollow spherical LiFePO$_4$ composites reported recently.

Surfactant	Size/Primary Particle Size	Initial Capacity (mAh g^{-1}, 0.1 C)	Cyclic Performance (mAh g^{-1}, 10 C)	Reference
CTAB	1~3 µm/100 nm	163	118	this work
-	200 nm	151	124	[44]
TRITON H-66	30 µm/100–200 nm	-	-	[28]
TRITON H-66	2 µm/100–400 nm	139	96	[27]
-	2 µm	158	101, 20 C, 2000th	[29]
CTAB	240 nm/30–50 nm	135	103	[38]
EDTMP	1–5 µm/20 nm	166	97, 20 C/80, 30 C	[45]
CTAB	420 nm	140	133	[32]

3. Materials and Methods

3.1. Synthesis Procedure

All chemicals were analytical grade. The LiFePO$_4$/C sample (LFP-A) was prepared by the solvothermal method in an autoclaved stainless steel reactor. The Li$_3$PO$_4$ white colloids were firstly precipitated from the precursors (LiOH, NH$_4$H$_2$PO$_4$) in EG medium (60% EG + 40% water in volume). Simultaneously, the FeSO$_4$ and CTAB were dissolved into EG solution in another container. After 1 h stirring, the mixture solution of FeSO$_4$ and CTAB was slowly added into the Li$_3$PO$_4$ suspension under mild magnetic stirring at room temperature for 15 min. The molar ratio of Li$^+$:PO$_4^{3-}$:Fe^{2+}: CTAB was 3:1:1:0.4. Then, the mixture was transferred into a stainless steel autoclave with Teflon lining. The autoclave was sealed and heated at 180 °C for 6 h. The carbon for the LiFePO$_4$ coating was sourced from a sucrose solution impregnating-drying-sintering procedure. The sintering was done at 650 °C for 2 h under 5 vol % H$_2$/Ar atmosphere with a heating rate of 5 °C min^{-1}. As a comparison, another sample named LFP-B was also synthesized via the same process excepting the absence of CTAB surfactant.

3.2. Characterization

The phase of the sample was analyzed via X-ray diffraction by a Rigaku D\max-2550 X-ray diffractometer (Rigaku, Tokyo, Japan) using Cu Kα_1 radiation, and the micro structure and morphology were characterized by field emission scanning electron microscopy (SEM, JEOL JSM6700F, Tokyo, Japan), and high-resolution transmission electron microscopy (HRTEM, JOEL JEM-2010F, Tokyo, Japan).

3.3. Electrochemical Performance

The electrochemical performances of the samples were evaluated using coin cells (Type 2016). The cathode was prepared by mixing the fabricated powder as active material (80 wt %), acetylene black (10 wt %) and poly(vinylidene fluoride) (PVDF) (10 wt %) dissolved in N-methyl-pyrrolidone (NMP). The slurry was cast onto an Al foil and dried at 80 °C for 12 h. Then the foil was cut into circular discs with a diameter of 1.5 cm by a precision disc cutter (MSK-T06, Hefei KE JING Materials Technology Co., Ltd., HeFei, China). A 1 M LiPF$_6$ solution dissolved in a mixture of ethylene carbonate (EC) and dimethyl carbonate (DMC) (EC:DMC = 1:1 in volume) was used as the electrolyte. The cells were assembled in an argon-filled glove box with lithium metal as the anode and microporous polypropylene sheet (Celgard 2400, Celgard, LLC, Charlotte, NC, USA) as a separator.

The cells were tested at different current densities of charge/discharge within the voltage range of 2.5–4.2 V using a LAND battery tester (Wuhan LAND Electronics Co. Ltd., Wuhan, China) at room temperature (25 °C). Cyclic voltammetry (CV) curves were tested at 0.1 mV s^{-1} within the range of 2.2–4.5 V at room temperature by an electrochemical workstation (CHI660E, Shanghai, China). Electrochemical impedance spectroscopy (EIS) was undertaken with an amplitude of 5 mV within the frequency range of 0.01 Hz to 100 kHz by using an Autolab PGSTAT302N (Metrohm, Herisau, Switzerland).

4. Conclusions

In summary, LiFePO$_4$/C composite in the form of hollow micro-spheres with a porous interior structure were synthesized through a solvothermal method with the assistance of CTAB surfactant dissolved in an EG solvothermal reaction medium. The addition of CTAB is responsible for the morphology change from spindles to hollow microspheres composed of nanoparticles. The as-obtained LiFePO$_4$/C hollow micro-spheres have a small distribution of diameters between 1 and 3 μm, with a ~300 nm thickness porous shell or wall composed of ~100 nm length nanoparticles. The porous shell/wall and loose interior of the hollow spheres make it easy to bring into contact with the electrolyte, facilitating fast, high-volume electronic and lithium ionic diffusion. Electrochemical measurements demonstrated that the hollow spherical LiFePO$_4$/C composite displayed excellent electrochemical performances: large reversible discharge capacity of 163 mAh g^{-1} at a current density of 0.1 C-rate (29.5 × 10^{-3} mA cm^{-2}), good rate capacity of 103 mAh g^{-1} at the 10 C rate, and superior capacity retention after various current density cycling. The high-scoring results can be attributed to the hollow spherical structure composed of nanosized particles, which facilitates fast electrochemical reaction kinetics and good structural stability. This may provide an effective strategy to produce other LiMPO$_4$ (M = Mn, Co, or Ni) hollow micro-spheres as cathode materials for lithium ion batteries.

Acknowledgments: This work is financially supported by National Natural Science Foundation of China (grant no. 51572166). The authors thank the programme for Professor of Special Appointment (Eastern Scholar) at Shanghai Institution of Higher Learning, and the Analysis and Research Center of Shanghai University for their technical support (XRD, D/MAX; SEM, JEOL JSM6700F; HRTEM, JOEL JEM-2010F).

Author Contributions: Yang Liu, Jieyu Zhang, Ying Li, and Yemin Hu conceived and designed the experiments, Yang Liu conducted the synthesis and performed the electrical measurements; Pengfei Hu performed the TEM measurements; Yang Liu and Shulei Chou analyzed the data of the respective experiments; Yang Liu, Wenxian Li, Mingyuan Zhu, Jieyu Zhang, and Guoxiu Wang discussed the data, Yang Liu wrote the paper with contributions from Wenxian Li and Guoxiu Wang; and all authors have given approval to the final version of the manuscript.

Conflicts of Interest: The authors declare no conflict of interest.

References

1. Padhi, A.K.; Nanjundaswamy, K.S.; Goodenough, J.B. Phospho-olivines as positive-electrode materials for rechargeable lithium batteries. *J. Electrochem. Soc.* **1997**, *144*, 1188–1194. [CrossRef]

2. Yamada, A.; Chung, S.C.; Hinokuma, K. Optimized LiFePO$_4$ for lithium battery cathodes. *J. Electrochem. Soc.* **2001**, *148*, A224–A229. [CrossRef]

3. Kang, B.; Ceder, G. Battery materials for ultrafast charging and discharging. *Nature* **2009**, *458*, 190–193. [CrossRef] [PubMed]

4. Padhi, A.K.; Nanjundaswamy, K.S.; Masquelier, C.; Okada, S.; Goodenough, J.B. Effect of structure on the Fe^{3+}/Fe^{2+} redox couple in iron phosphates. *J. Electrochem. Soc.* **1997**, *144*, 1609–1613. [CrossRef]

5. Chung, S.Y.; Bloking, J.T.; Chiang, Y.M. Electronically conductive phospho-olivines as lithium storage electrodes. *Nat. Mater.* **2002**, *1*, 123–128. [CrossRef] [PubMed]

6. Wang, Y.; Wang, Y.; Hosono, E.; Wang, K.; Zhou, H. The design of a LiFePO$_4$/carbon nanocomposite with a core-shell structure and its synthesis by an in situ polymerization restriction metho. *Angew. Chem.* **2008**, *47*, 7461–7465. [CrossRef] [PubMed]

7. Xiang, H.; Zhang, D.; Jin, Y.; Chen, C.; Wu, J.; Wang, H. Hydrothermal synthesis of ultra-thin LiFePO$_4$ platelets for Li-ion batteries. *J. Mater. Sci.* **2011**, *46*, 4906–4912. [CrossRef]

8. Yang, J.; Wang, J.; Wang, D.; Li, X.; Geng, D.; Liang, G.; Gauthier, M.; Li, R.; Sun, X. 3D porous LiFePO$_4$/graphene hybrid cathodes with enhanced performance for Li-ion batteries. *J. Power Sources* **2012**, *208*, 340–344. [CrossRef]

9. Gaberscek, M.; Dominko, R.; Jamnik, J. Is small particle size more important than carbon coating? An example study on LiFePO$_4$ cathodes. *Electrochem. Commun.* **2007**, *9*, 2778–2783. [CrossRef]

10. Julien, C.M.; Mauger, A.; Zaghib, K. Surface effects on electrochemical properties of nano-sized LiFePO$_4$. *J. Mater. Chem.* **2011**, *21*, 9955–9968. [CrossRef]

11. Wang, Y.; Sun, B.; Park, J.; Kim, W.S.; Kim, H.S.; Wang, G.X. Morphology control and electrochemical properties of nanosize LiFePO$_4$ cathode material synthesized bi co-precipitation combined with in situ polymerization. *J. Alloys Compd.* **2011**, *509*, 1040–1044. [CrossRef]

12. Drezen, T.; Kwon, N.H.; Bowen, P.; Teerlinck, I.; Isono, M.; Exnar, I. Effect of particle size on LiMnPO$_4$ cathodes. *J. Power Sources* **2007**, *174*, 949–953. [CrossRef]

13. Dimesso, L.; Forster, C.; Jaegermann, W.; Khanderi, J.P.; Tempel, H.; Popp, A.; Engstler, J.; Schneider, J.J.; Sarapulova, A.; Mikhailova, D.L.; et al. Developments in nanostructured LiMPO$_4$ (M = Fe, Co, Ni, Mn) composites based on three dimensional carbon architecture. *Chem. Soc. Rev.* **2012**, *41*, 5068–5080. [CrossRef] [PubMed]

14. Wang, J.; Sun, X. Understanding and recent development of carbon coating on LiFePO$_4$ cathode materials for lithium-ion batteries. *Energy Environ. Sci.* **2012**, *5*, 5163–5185. [CrossRef]

15. Wang, G.X.; Yang, L.; Chen, Y.; Wang, J.Z.; Bewlay, S.; Liu, H.K. An investigation of polypyrrolr-LiFePO$_4$ composite cathode materials for lithium-ion batteries. *Electrochim. Acta* **2005**, *50*, 4649–4654. [CrossRef]

16. Wu, Y.; Wen, Z.; Li, J. Hierarchical carbon-coated LiFePO$_4$ nanoplate microspheres with high electrochemical performance for Li-ion batteries. *Adv. Mater.* **2011**, *23*, 1126–1129. [CrossRef] [PubMed]

17. Liu, H.; Cao, Q.; Fu, L.J.; Li, C.; Wu, Y.P.; Wu, H.Q. Doping effects of zinc on LiFePO$_4$ cathode material for lithium ion batteries. *Electrochem. Commun.* **2006**, *8*, 1553–1557. [CrossRef]

18. Yang, J.; Bai, Y.; Qing, C.; Zhang, W. Electrochemical performances of Co-doped LiFePO$_4$/C obtained by hydrothermal method. *J. Alloys Compd.* **2011**, *509*, 9010–9014. [CrossRef]

19. Shenouda, A.Y.; Liu, H.K. Studies on electrochemical behavior of zinc-doped LiFePO$_4$ for lithium battery positive electrode. *J. Alloys Compd.* **2009**, *477*, 498–503. [CrossRef]

20. Jin, Y.; Yang, C.P.; Rui, X.H.; Cheng, T.; Chen, C.H. V$_2$O$_3$ modified LiFePO$_4$/C composite with improved electrochemical performance. *J. Power Sources* **2011**, *196*, 5623–5630. [CrossRef]

21. Hu, Y.; Yao, J.; Zhao, Z.; Zhu, M.; Li, Y.; Jin, H.; Zhao, H.; Wang, J. ZnO-doped LiFePO$_4$ cathode material for lithium-ion battery fabricated by hydrothermal method. *Mater. Chem. Phys.* **2013**, *141*, 835–841. [CrossRef]

22. Arico, A.S.; Bruce, P.; Scrosati, B.; Tarascon, J.M.; Schalkwijk, W.V. Nanostructured materials for advanced energy conversion and storage devices. *Nat. Mater.* **2005**, *4*, 366–377. [CrossRef] [PubMed]

23. Cao, J.; Qu, Y.; Guo, R. La$_{0.6}$Sr$_{0.4}$CoO$_{3-\delta}$ modified LiFePO$_4$/C composite cathodes with improved electrochemical performances. *Electrochim. Acta* **2012**, *67*, 152–158. [CrossRef]

24. Salah, A.A.; Mauger, A.; Zaghib, K.; Goodenough, J.B.; Ravet, N.; Gauthier, M.; Gendron, F.; Julien, C.M. Reduction Fe^{3+} of impurities in LiFePO$_4$ from pyrolysis of organic precursor used for carbon deposition. *J. Electrochem. Soc.* **2006**, *153*, A1692–A1701. [CrossRef]

25. Li, H.; Zhou, H. Enhancing the performances of Li-ion batteries by carbon-coating: Present and future. *Chem. Commun.* **2012**, *48*, 1201–1217. [CrossRef] [PubMed]

26. Lee, K.S.; Myung, S.T.; Sun, Y.K. Microwave synthesis of spherical Li[Ni$_{0.4}$Co$_{0.2}$Mn$_{0.4}$]O$_2$ powders as a positive electrode material for lithium batteries. *Chem. Mater.* **2007**, *19*, 2727–2729. [CrossRef]

27. Park, Y.; Roh, K.C.; Shin, W.; Lee, J.W. Novel morphology-controlled synthesis of homogeneous LiFePO$_4$ for Li-ion batteries using an organic phosphate source. *RSC Adv.* **2013**, *3*, 14263–14266. [CrossRef]

28. Park, Y.; Shin, W.; Lee, J.W. Synthesis of hollow spherical LiFePO$_4$ by a novel route using organic phosphate. *CrystEngComm* **2012**, *14*, 4612–4617. [CrossRef]

29. Yang, S.; Hu, M.; Xi, L.; Ma, R.; Dong, Y.; Chung, C.Y. Solvothermal synthesis of monodisperse LiFePO$_4$ micro hollow spheres as high performance cathode material for lithium ion batteries. *ACS Appl. Mater. Interfaces* **2013**, *5*, 8961–8967. [CrossRef] [PubMed]

30. Ren, Y.; Bruce, P.G. Mesoporous LiFePO$_4$ as a cathode material for rechargeable lithium ion batteries. *Electrochem. Commun.* **2012**, *17*, 60–62. [CrossRef]

31. Huang, Z.D.; Oh, S.W.; He, Y.B.; Zhang, B.; Yang, Y.; Mai, Y.W.; Kim, J.K. Porous C-LiFePO$_4$-C composite microspheres with a hierarchical conductive architecture as a high performance cathode for lithium ion batteries. *J. Mater. Chem.* **2012**, *22*, 19643–19645. [CrossRef]

32. Lee, M.H.; Kim, J.Y.; Song, H.K. A hollow sphere secondary structure of LiFePO$_4$ nanoparticles. *Chem. Commun.* **2010**, *46*, 6795–6797. [CrossRef] [PubMed]

33. Lee, M.H.; Kim, T.H.; Kim, Y.S.; Park, J.S.; Song, H.K. Optimized evolution of a secondary structure of LiFePO$_4$: Balancing between shape and impurities. *J. Mater. Chem.* **2012**, *22*, 8228–8234. [CrossRef]

34. Cho, M.Y.; Kim, K.B.; Lee, J.W.; Kim, H.; Kim, H.; Kang, K.; Roh, K.C. Defect-free solvothermally assisted synthesis of microspherical mesoporous LiFePO$_4$/C. *RSC Adv.* **2013**, *3*, 3421–3427. [CrossRef]

35. Sun, C.; Rajasekhara, S.; Goodenough, J.B.; Zhou, F. Monodisperse porous LiFePO$_4$ microspheres for a high power Li-ion battery cathode. *J. Am. Chem. Soc.* **2011**, *133*, 2132–2135. [CrossRef] [PubMed]

36. Su, J.; Wu, X.L.; Yang, C.P.; Lee, J.S.; Kim, J.; Guo, Y.G. Self-assembled LiFePO$_4$/C nano/microspheres by using phytic acid as phosphorus source. *J. Phys. Chem. C* **2012**, *116*, 5019–5024. [CrossRef]

37. Yao, J.; Tjandra, W.; Chen, Y.Z.; Tam, K.C.; Ma, J.; Soh, B. Hydroxyapatite nanostructure material derived using cationic surfactant as a template. *J. Mater. Chem.* **2003**, *13*, 3053–3057. [CrossRef]

38. Lee, M.H.; Kim, T.H.; Kim, Y.S.; Song, H.K. Precipitation revisited: Shape control of LiFePO$_4$ nanoparticles by combinatorial precipitation. *J. Phys. Chem. C* **2011**, *115*, 12255–12259. [CrossRef]

39. Belharouak, I.; Johnson, C.; Amine, K. Synthesis and electrochemical analysis of vapor-deposited carbon-coated LiFePO$_4$. *Electrochem. Commun.* **2005**, *7*, 983–988. [CrossRef]

40. Zhang, Y.; Wang, W.; Li, P.; Fu, Y.; Ma, X. A simple solvothermal route to synthesize graphene-modified LiFePO$_4$ cathode for high power lithium ion batteries. *J. Power Sources* **2012**, *210*, 47–53. [CrossRef]

41. Dathar, G.K. P.; Sheppard, D.; Stevenson, K.J.; Henkelman, G. Calculations of Li-ion diffusion in olivine phosphates. *Chem. Mater.* **2011**, *23*, 4032–4037. [CrossRef]

42. Jin, B.; Jin, E.M.; Park, K.H.; Gu, H.B. Electrochemical properties of LiFePO$_4$-multiwalled carbon nanotubes composite cathode materials for lithium polymer battery. *Electrochem. Commun.* **2008**, *10*, 1537–1540. [CrossRef]

43. Meng, H.; Zhou, P.; Zhang, Z.; Tao, Z.; Chen, J. Preparation and characterization of LiNi$_{0.8}$Co$_{0.15}$Al$_{0.05}$O$_2$ with high cycling stability by using AlO$_2$$^-$ as Al source. *Ceram. Int.* **2017**, *43*, 3885–3892. [CrossRef]

44. Bi, J.; Zhang, T.; Wang, K.; Zhong, B.; Luo, G. Controllable synthesis of Li$_3$PO$_4$ hollow nanospheres for the preparation of high performance LiFePO$_4$ cathode material. *Particuology* **2016**, *24*, 142–150. [CrossRef]

45. Xiao, P.; Lai, M.O.; Lu, L. Hollow microspherical LiFePO$_4$/C synthesized from a novel multidentate phosphonate complexing agent. *RSC Adv.* **2013**, *3*, 5127–5130. [CrossRef]

nanomaterials

MDPI

Article

3D Nanoporous Anodic Alumina Structures for Sustained Drug Release

Maria Porta-i-Batalla, Elisabet Xifré-Pérez, Chris Eckstein, Josep Ferré-Borrull and Lluis F. Marsal *

Departament d'Enginyeria Electrònica, Elèctrica i Automàtica, ETSE, Universitat Rovira i Virgili, Avda. Països Catalans 26, 43007 Tarragona, Spain; maria.porta@urv.cat (M.P.-i.-B.); elisabet.xifre@urv.cat (E.X.-P.); chris.eckstein@urv.cat (C.E.); josep.ferre@urv.cat (J.F.-B.)
* Correspondence: lluis.marsal@urv.cat; Tel.: +34-977-55-96-25

Received: 21 July 2017; Accepted: 14 August 2017; Published: 21 August 2017

Abstract: The use of nanoporous anodic alumina (NAA) for the development of drug delivery systems has gained much attention in recent years. The release of drugs loaded inside NAA pores is complex and depends on the morphology of the pores. In this study, NAA, with different three-dimensional (3D) pore structures (cylindrical pores with several pore diameters, multilayered nanofunnels, and multilayered inverted funnels) were fabricated, and their respective drug delivery rates were studied and modeled using doxorubicin as a model drug. The obtained results reveal optimal modeling of all 3D pore structures, differentiating two drug release stages. Thus, an initial short-term and a sustained long-term release were successfully modeled by the Higuchi and the Korsmeyer–Peppas equations, respectively. This study demonstrates the influence of pore geometries on drug release rates, and further presents a sustained long-term drug release that exceeds 60 days without an undesired initial burst.

Keywords: drug delivery; nanoporous anodic alumina; complex pore geometry; release rate; nanotechnology

1. Introduction

A new generation of local drug release platforms with sustained and complex release profiles for reduced therapeutic doses has emerged to overcome the disadvantages of conventional treatments—generally oral or intravenous administrations—with considerable adverse effects [1,2].

These local drug release platforms face important challenges to ensure efficient therapy: (i) efficient loading of drugs, (ii) sustained delivery, (iii) avoiding the 'burst effect' (high dose release within the first minutes), (iv) material stability (avoiding degradation), and (v) the possibility to chemically modify their surface for a selective release [3–8]. Many types of materials are currently used for the development of these new drug delivery platforms, such as polymers [9], hydrogels [10,11], iron oxide [12], graphene [13], porous silicon [7], and mesoporous silica [14]. However, most of these existing carrier materials rapidly degrade at physiological pH or/and show poor drug loading with drugs mainly attaching to external surfaces and leading to an intense initial "burst" release.

Porous materials have attracted great interest for the development of controlled drug delivery platforms because of their high effective surface area and tunable pore size [15]. The pore geometry is one of the main determining factors of the total drug load entering the pores and the release profile. Three-dimensional (3D) pore structures, with intricate pore geometries and increasing surface area are promising platform designs for sustained drug release. However, their fabrication can be expensive and complex.

Nanoporous anodic alumina (NAA), readily and cost-effectively fabricated by electrochemical anodization, permits obtaining elaborate and reproducible 3D pore geometries. The many physical and chemical properties of NAA make this material a versatile and interesting platform for controlled drug release. NAA has a highly ordered pore distribution, and its well-known electrochemical fabrication techniques allow for the precise control of pore diameter, interpore distance, pore length, and pore geometry [16–18]. NAA is highly stable at physiological pH, and has been successfully used in a wide array of medical and biological applications like orthopedic prosthetics, dental and coronary stents, cell culture scaffolds, immunoisolation devices, and biomolecular filtration [2,19–21].

Its high effective surface area makes NAA an ideal material for drug delivery applications providing pores as nanocontainers with regular and controlled structural features for loading active agents like drugs or molecules [22,23]. Moreover, the surface of NAA can be functionalized to be selective for specific molecules and covered with biodegradable, chemical, or pH responsive agents to trigger and regulate the release [24–26].

Although drug release from nanoporous coatings has already been studied, there is a lack of understanding of the release kinetics from NAA platforms with complex pore geometries and the dynamics governing them [27].

In this work the drug release kinetics for simple and complex NAA pore structures is investigated. Pores with straight walls and 3D pore structures with multilayered funnel and inverted funnel geometries are fabricated by electrochemical anodization, resulting in complex NAA platforms for drug delivery. Using the chemotherapeutic Doxorubicin, the drug release of these different pore geometries is studied, and the release mechanism is modeled by mathematical expressions.

2. Materials and Methods

2.1. Fabrication of NAA Structures

All NAA porous structures were prepared by electrochemical anodization of high purity (99.999%) aluminum plates (Goodfellow, Huntingdon, UK) in phosphoric acid electrolyte. The aluminum plates were initially degreased with acetone and ethanol to eliminate organic impurities and electropolished in a mixture of perchloric acid and ethanol 1:4 (v/v) at a constant voltage of 20 V for 6 min. To suppress breakdown effects and to enable uniform oxide film growth under hard anodization conditions (194 V in phosphoric acid at −5 °C), a protective layer was pre-anodized at a lower voltage (174 V in phosphoric acid) for 180 min [28]. Subsequently, the voltage was ramped up to 194 V at a constant rate of 0.05 V/s, and anodized for 20 h. After this first anodization step, the formed NAA layer was removed by wet chemical etching in a mixture of phosphoric acid (0.4 M), and chromic acid (0.2 M) at 70 °C for 4 h, resulting in a hexagonally-ordered pattern of the aluminum surface [29,30].

For all pore structures, the subsequent anodization steps were performed under hard anodization conditions, and the length of the pores was accurately controlled via the total charge.

A single hard anodization step was performed to obtain straight pores (SP) with a uniform pore diameter from top to bottom (Figure 1a). The length of the pores of all SP is 30 μm. To widen the pores, wet chemical etching with aqueous solution of 5% H_3PO_4 was performed for 0 (SP1), 45 (SP2), 90 (SP3), and 120 min (SP4).

Normal Funnels (NF) were produced by a sequential combination of hard anodization and pore widening steps, and labeled according to the final number of layers. NF2 consist of a 15 μm thick top layer, widened in 5% H_3PO_4 for 90 min, and a 15 μm thick bottom layer. Similarly, NF3 consist of a 10 μm thick top layer widened for a total of 90 min (2 × 45 min), a middle layer of 10 μm widened for 45 min, and a bottom layer of 10 μm (Figure 1b).

The fabrication of Inverted Funnels (IF) required a thermal treatment at 250 °C and 500 °C to change the crystallographic phase of the alumina. The inverted funnels were labeled IF2 and IF3 according to their respective total number of layers (Figure 1c). IF2 consisted of a top layer of 15 μm, followed by a thermal treatment at 500 °C. A subsequent anodization step added a 15 μm thick bottom layer. Similarly, IF3 consisted of 3 layers, each with a thickness of 10 μm. After the anodization of the top and middle layer a thermal treatment of 500 °C and 250 °C was applied, respectively. In a final wet chemical etching step, the pores of IF2 and IF3 were widened for 2 h.

All NAA platforms were cut with a circular cutter to obtain samples with the same diameter (4 mm). All the experiments were conducted in triplicates.

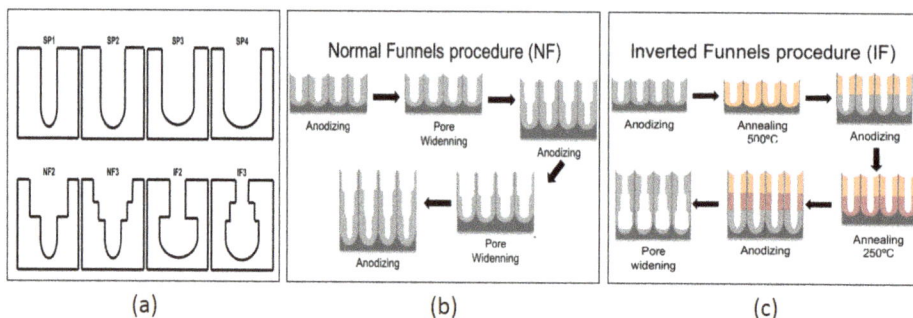

Figure 1. (**a**) Schematic illustration and labeling of the different pore geometries; (**b**) Schematics of the fabrication procedures for normal funnels; and (**c**) inverted funnels.

2.2. Characterization of NAA Structures

All NAA structures were characterized by Environmental Scanning Electron Microscopy (ESEM, FEI Quanta 600, FEI Co., Hillsboro, OR, USA). The wet chemical etch rate during the pore widening steps was estimated for samples with and without a 500 °C thermal treatment. For the calibration of the pore widening process, ESEM images were taken in 15 min etching intervals, and the pore diameter was estimated using a standard image processing package (ImageJ, version 1.51p, public domain program developed at the RSB of the NIH, Bethesda, Maryland, MD, USA) (Figure S1).

2.3. Drug Loading and Release

Doxorubicin (DOX), a self-fluorescent chemotherapeutic agent, was selected as a model drug. Drug loading into NAA pores was performed through capillary action by immersing NAA into a DOX solution of 1 mg/mL. The suspension was stirred overnight in the dark with the NAA structures immersed. Subsequently, samples were washed with deionized water to remove any residual drug molecules on the surface of the sample and dried at an ambient temperature.

The release studies were performed in vitro using phosphate-buffered saline (PBS), which is commonly employed to simulate in vivo conditions for drug release. DOX release was estimated by directly measuring the photoluminescence of the release medium. This in situ measurement process is ideal to understand the release kinetics and the short-term release effect since it allows for the fast and frequent collection of release data. Samples were immersed in 0.5 mL of PBS which was renewed after every measurement. The fluorescence of the buffer solution was measured at regular time intervals at room temperature using a fluorescence spectrophotometer from Photon Technology International Inc. (Birmingham, NJ, USA), with an Xe lamp as the excitation light source, an excitation wavelength of 480 nm and an emission wavelength of 590 nm. The drug release was monitored by DOX photoluminescence over 65 days. The fluorescence intensities were converted to the corresponding concentrations using a calibration curve. All the drug release measurements were taken in triplicates for every pore structure and statistical analysis was performed.

3. Results and Discussion

Normal Funnels (NF) with two and three layers of different pore diameters were successfully fabricated. ESEM cross-section images of NF show straight pore growth with no discontinuities (i.e., occluded pores) despite the interruption of the anodization process between layers (Figure 2). The transition between adjacent layers of different pore sizes is smooth and shows a conical shape. The pore diameter and the thickness of each NF layer was estimated from ESEM images and summarized in Table S1.

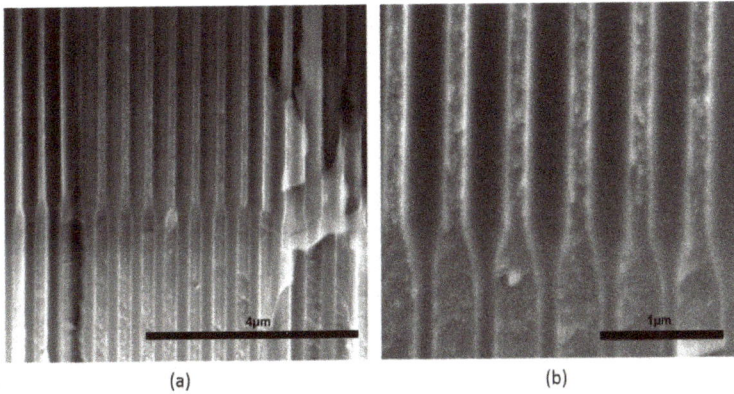

Figure 2. Cross-section Environmental Scanning Electron Microscopy (ESEM) images of a two layered normal funnel structure (NF2) illustrating (**a**) the uniform pore growth, and (**b**) the pore transition.

Inverted Funnels (IF) were also successfully fabricated, and their ESEM cross-section images are shown in Figure 3 (IF2) and Figure 4 (IF3). For both IF, parallel and perpendicular growth of the pores as well as conic and clear transitions between adjacent layers of different pore diameters can be observed. High magnification images show that these conic transitions between layers are more abrupt for IF than for NF. IF, NF, and SP were anodized to a total length of 30 μm for better comparability (Table S1).

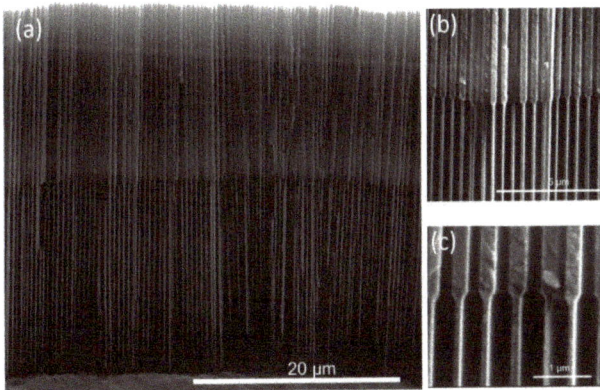

Figure 3. Cross section ESEM images of a two layered inverted funnel structure (IF2): (**a**) entire bilayer, (**b**) detail of the uniform pore growth, and (**c**) detail of the pore transition.

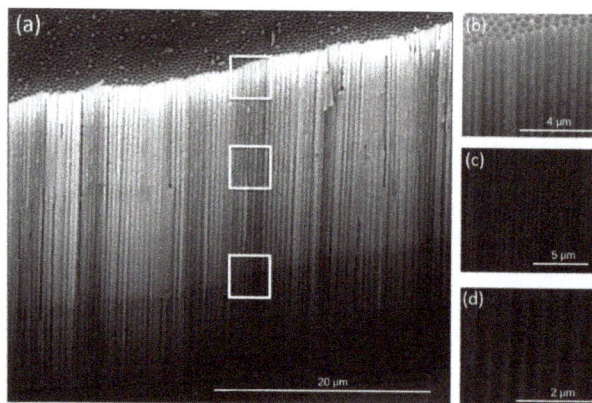

Figure 4. Cross section ESEM images of a three layered inverted funnel structure (IF3): (**a**) entire length of the inverted funnels, and (**b–d**) magnifications of the framed areas from top to bottom.

The two distinct pore diameters of the IF2 top and bottom layers suggest that the crystallographic structure of the top layer was successfully modified by the temperature treatment (500 °C). The same clear transitions between layers are observed in IF3 samples, demonstrating that the intermediate annealing temperature of 250 °C results in an intermediate pore widening rate.

The effect of the temperature treatment on the pore widening process during IF fabrication was further assessed, and a calibration of the pore widening rates was determined. The pore diameters were estimated from ESEM images taken after consecutive 15 min pore widening steps for samples thermally treated at 500 °C, and for untreated samples. Figure S1 reveals that the thermally treated samples have a slower pore widening rate than untreated samples. The alumina matrix of untreated samples remained intact for up to 2 h of etching, but started to deteriorate after 2.5 h, and fully collapsed after 3 h due to over-etched pore walls. Consequently, to preserve full structural integrity of the thermally untreated samples, the pore widening was terminated after a maximum of 2 h. In contrast, the pore structure of thermally treated samples remains intact even after 3 h of etching.

Figure S2 shows the estimated pore diameter as a function of pore widening time, revealing higher etch rates for the untreated samples (2.5 nm/min) than for the thermally treated samples (1.2 nm/min). These are brought about by the thermal annealing, which increases the crystallinity, and consequently, the stability of the alumina, and also further promotes anion diffusion [31]. We further notice that the pore diameter linearly increases with the pore widening time until an inflection point, where the pore widening rate decreases considerably. This inflection point corresponds to an interface separating an outer region with concentrated anionic species from an inner region composed of pure alumina. The anion contaminated region is easily removed by the pore widening, whereas the region of pure alumina is more resistant to pore widening [32,33]. The inflection point is observed after 90 min and 150 min of pore widening for non-treated samples and thermally treated samples, respectively.

Figure 5a shows the pore diameter of the top layers of all the fabricated NAA structures. Increasing pore diameters of samples SP1–SP4 are directly related to the increasing widening time intervals (Table S1). NF2 and NF3 show top pore diameters similar to SP3 (around 300 nm), and IF2 and IF3 top diameters are similar to that of SP4 (around 200 nm). These similarities allow for studying the influence of the pore geometry on the release kinetics. Figure 5b shows the total volume of all of the samples. The pore volumes of SP1–SP4 are clearly related to the pore diameters, whereas the pore volumes of the layered samples NF and IF depend on the complex geometries of the pores.

The relationship between the total amount of drug load and volume is shown in Figure 6. The total drug load of straight pores is linearly proportional to the pore volume, as indicated by the linear regression for SP1–SP4 (dotted line). Similarly, the total drug load of NF follows the trend of SP samples, though, with an increasing diversion from this trend with more funnel layers. Interestingly, the IF samples seem to hold a higher drug load per pore volume when compared to SP, indicating an influence of the pore geometry on the total drug load. This can be explained by the contour of the IF pores which encompasses small pore diameters of the top layer, wider pore diameters of deeper layers and a sharp elbow-like transitions between the adjacent layers (Figure 3c). This intricate geometry retains a higher total drug load within the pores, making IF structures more efficient for drug loading than SP structures.

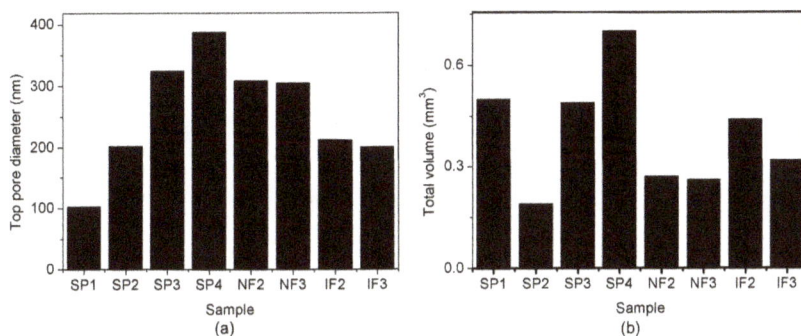

Figure 5. (**a**) Top pore diameter, and (**b**) total volume of the different samples.

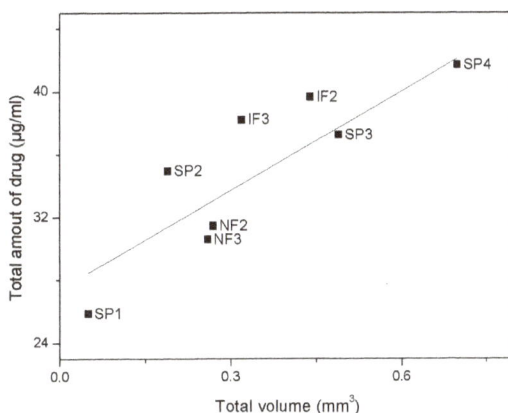

Figure 6. Total drug load within the pores versus the respective total volume for various nanoporous anodic alumina (NAA) structures. Linear regression for straight pores (SP) samples (dotted line).

The influence of the pore geometry on short and long-term drug release has been studied for all fabricated NAA structures. Figure 7 and Figure S3 show the drug release response for the first 8 h, and for the complete release time of 1512 h (63 days). All of the NAA structures presented in this work can be considered as sustained drug delivery platforms due to very long drug release times. Interestingly, these NAA structures do not present a high initial drug release burst in the first minutes, in contrast to most drug delivery platforms in the literature [34–36]. The absence of an initial release burst indicates that the drug delivery from these NAA platforms is more constant in time, and prevents an undesired high initial dosage. Both the sustained delivery, and the absence of initial burst are relevant and differentiating properties and address two of the main challenges of localized drug delivery.

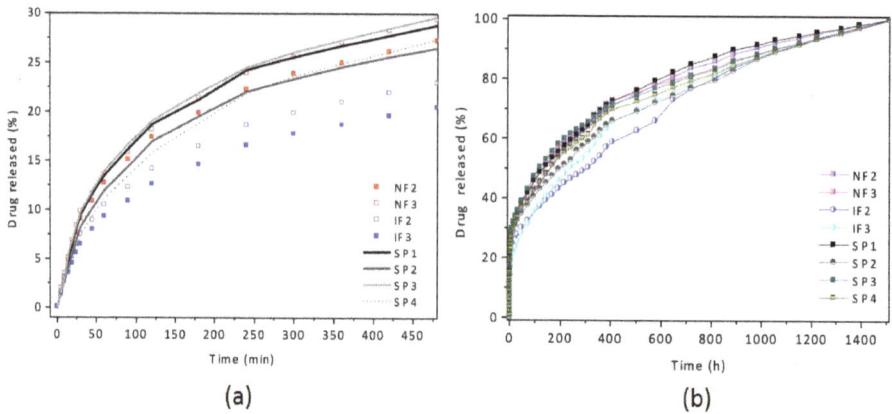

Figure 7. Cumulative drug release of different pore structures: (**a**) short-term and (**b**) long-term.

The drug release profile of all the pore geometries can be described by distinguishing two phases: (i) a short-term release with a higher release rate within the first 8 h, and (ii) a slow and sustained release where almost the entire drug load is delivered from the NAA after 63 days.

Most of the NAA structures presented here release only around 25–30% of the total drug load during the short-term release, which is very low compared to the 80% and above of most conventional structures in the literature [37,38]. Generally, the initial release is attributed to the fast diffusion of drug molecules residing on the NAA surface, rather than the diffusion of molecules attached to the walls within the pores. Here, this low short-term release indicates that most of the drug was loaded inside the pores during the incubation period.

The pore geometries were found to influence the short-term release rates. The 200 nm pore diameter of the top layer of IF2 and IF3 is similar to SP2, however, their release rates are considerably lower than SP2. The pore opening of IF acts like a bottleneck for the infiltrating medium and the eluting drug, hindering the circulation of the medium inside the pore and slowing down the diffusion of the drug out of the alumina. This effect augments with increasing IF layers: the release rate of IF3 is lower than that of IF2.

On the contrary, the NF and SP3, with very close top pore diameters of around 310 nm, present similar release profiles, the geometry of NF does not hinder the circulation of the medium inside of the pores.

The experimental data for short- and long-term release were modeled to measure the kinetics and establish the mechanism of DOX release. For short-term release, the experimental data was modeled using a variation of the Higuchi equation [9,39–41]:

$$M_t = M_0 + K\sqrt{t} \tag{1}$$

where M_t is the cumulative release at time t, M_0 is the intercept value at $t = 0$ and, K is the release constant that indicates the release velocity.

The fitting of the experimental data for all pore geometries with Equation (1) is presented in Figure 8, where the cumulative DOX release is plotted against the square root of time during the short-term release. The fitting is in very good agreement with the experimental data for all pore geometries, demonstrating that the drug kinetics can be approximated by the square root of time. Table 1 shows the fitting parameters for each pore structure and Figure 9 depicts their release constant K depending on the top pore diameter and volume.

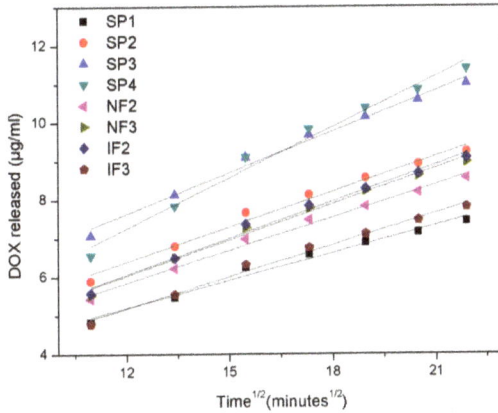

Figure 8. Short-term release data (symbols) and fitting (lines) using the Higuchi equation.

Table 1. Higuchi equation fitting parameters.

Sample Name	Release Constant (K)	Intercept (M_0)
SP1	0.24 ± 0.02	2.34 ± 0.26
SP2	0.30 ± 0.02	2.75 ± 0.29
SP3	0.35 ± 0.02	3.38 ± 0.29
SP4	0.44 ± 0.02	2.01 ± 0.43
NF2	0.28 ± 0.01	2.45 ± 0.21
NF3	0.31 ± 0.01	2.31 ± 0.25
IF2	0.32 ± 0.01	2.28 ± 0.25
IF3	0.27 ± 0.01	1.89 ± 0.20

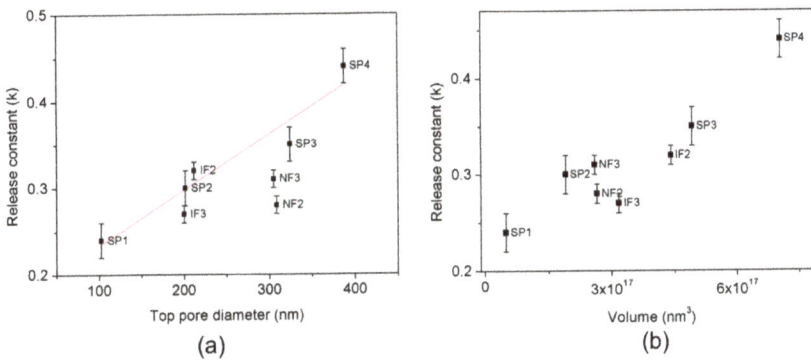

Figure 9. Release constant K during the short-term release versus (**a**) top pore diameter and, (**b**) volume. The red line corresponds to the linear fitting.

During the short-term release, the release constant K for straight pore structures (SP1–SP4) is linearly proportional to the pore diameter, following the equation:

$$K = 0.168 + 6.46 \times 10^{-4} \cdot D_p \tag{2}$$

where K is the release constant in $(\mu g/mL)/min^{1/2}$ and D_p is the pore diameter in nanometers.

IF are the structures with the highest load efficiency, as they retain a higher quantity of drug inside of the pores than SP and NF, with the same volume or top pore diameter. During short-term release, IF structures show a lower release constant K than SP structures with the same top pore diameter. However, if the specific application requires a higher release rate, SP structures are favorable. Even though SP and NF present similar release rates, the fabrication of SP structures is not as complex as the fabrication of NF.

For the long-term release the experimental data was modeled with the Korsmeyer–Peppas equation [6,38,42]:

$$M_t = M_{t0} \left(\frac{t}{t_0} \right)^n \tag{3}$$

where M_t is the quantity of drug released at time t, M_{t0} is the amount of drug released at the reference time t_0 (day 1), t is time in days, and n is the release parameter related to the release rate.

Table 2 shows the values of these parameters fitted for the release of all the pore structures. The release rate was calculated with the first derivative of Equation (3) [3,22]. Figure 10 shows a good agreement between the experimental data and the fitting modeled with Equation (3).

Table 2. Korsmeyer–Peppas equation fitting parameters.

Sample Name	M_{t0} (µg/mL)	n	Release Rate
SP1	8.43 ± 0.14	0.27 ± 0.01	2.31
SP2	9.93 ± 0.21	0.30 ± 0.001	2.94
SP3	12.49 ± 0.17	0.26 ± 0.00	3.27
SP4	12.55 ± 0.22	0.28 ± 0.01	3.60
NF2	9.70 ± 0.16	0.29 ± 0.00	2.78
NF3	10.14 ± 0.15	0.27 ± 0.00	2.69
IF2	9.00 ± 0.33	0.35 ± 0.01	3.16
IF3	8.91 ± 0.18	0.35 ± 0.01	3.13

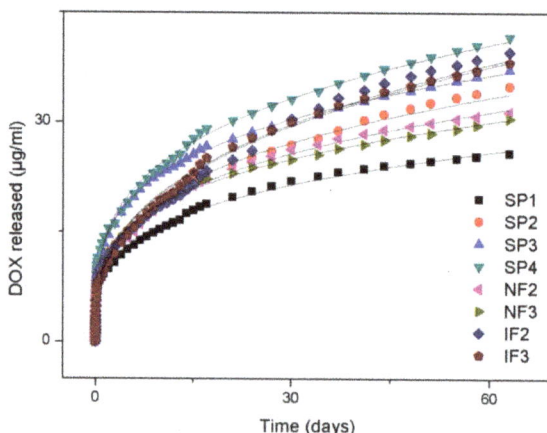

Figure 10. Long-term release experimental data (symbols) and fitting using the Korsmeyer–Peppas equation (lines).

For the long-term release, a linear relation between the release rate and the pore diameter is observed for SP (Figure 11):

$$\text{Release rate} = 1.95 + 0.004D_p \tag{4}$$

where D_p is the top pore diameter.

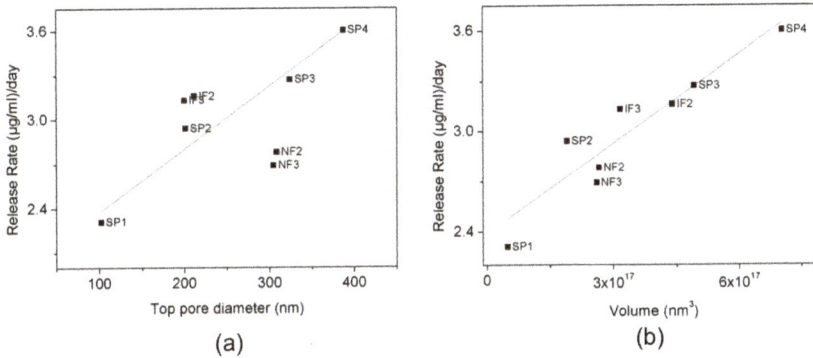

Figure 11. Release rate versus (**a**) top pore diameter, and (**b**) volume, and linear fittings (black straight line).

NF and SP3 have a similar top pore diameter and release rates. IF have slightly higher release rates than SP2, which is explained by the quantity of drug remaining inside the pores. During the short-term release, NF and SP delivered a higher part of their load than IF as their release rates were higher. Due to this, and the fact that drug loads were completely released from all the structures after 63 days, IF released greater loads during days 8–63 than the other structures, and therefore their release rate is slightly higher.

4. Conclusions

In this study, NAA platforms with a variety of 3D pore morphologies were fabricated, loaded with DOX, and the release mechanism was modeled by mathematical expressions. Besides, the influence of the pore geometry on the drug release kinetics was assessed.

The release profiles for the studied pore geometries revealed two interesting and promising properties: (i) very long drug release times determining the presented NAA as sustained drug delivery platforms, and (ii) a constant drug delivery free of an initial release burst to prevent undesired high initial drug delivery dosages. These findings are advancements to two of the main challenges of current platforms for advanced drug delivery systems.

The obtained results reveal that the pore geometry influences the total drug load within the pores. IF retain a higher quantity of drug inside the pores than SP and NF with the same volume or top pore diameter. The pore geometry also influences the release kinetics. During the short-term release, IF showed lower release rates than SP with the same top pore diameter.

Moreover, the dynamics of the release of all the pore structures were successfully modeled, and two different release regimes were differentiated: a short-term and a long-term release. The short-term release (first 8 h) was modeled by the Higuchi model, whereas for the long-term release the Korsmeyer–Peppas equations were used.

Supplementary Materials: The following are available online at http://www.mdpi.com/2079-4991/7/8/227/s1. Figure S1: Pore widening progress for samples with and without temperature treatment; Figure S2: Pore widening calibration for samples with and without thermal treatment; Figure S3: Cumulative drug release of different pore structures; Table S1: Average dimensions of the pore structures.

Acknowledgments: This work was supported by the Spanish Ministry of Economy and competitiveness (MINECO) under grant number TEC2015-71324-R (MINECO/FEDER, UE), the Catalan authority under project AGAUR 2014 SGR 1344, and ICREA 2014 under the ICREA Academia Award.

Author Contributions: M.P.B. carried out the experiments. L.F.M. conceived and designed the experiments and supervised the study. All authors analyzed the data, contributed to scientific discussion and wrote the manuscript.

Conflicts of Interest: The authors declare no conflict of interest.

References

1. Allen, T.M.; Cullis, P.R. Drug delivery systems: Entering the mainstream. *Science* **2004**, *303*, 1818–1822. [CrossRef] [PubMed]
2. Losic, D.; Simovic, S. Self-ordered nanopore and nanotube platforms for drug delivery applications. *Expert Opin. Drug Deliv.* **2009**, *6*, 1363–1381. [CrossRef] [PubMed]
3. Yao, F.; Weiyuan, J.K. Drug Release Kinetics and Transport Mechanisms of Nondegradable and Degradable Polymeric Delivery Systems. *Expert Opin. Drug Deliv.* **2011**, *7*, 429–444.
4. Gultepe, E.; Nagesha, D.; Casse, B.D.F.; Banyal, R.; Fitchorov, T.; Karma, A.; Amiji, M.; Sridhar, S. Sustained drug release from non-eroding nanoporous templates. *Small* **2010**, *6*, 213–216. [CrossRef] [PubMed]
5. Fornell, J.; Soriano, J.; Guerrero, M.; Sirvent, J.; Ferran-Marqués, M.; Ibáñez, E.; Barrios, L.; Baró, M.D.; Suriñach, S.; Nogués, C.; et al. Biodegradable FeMnSi Sputter-Coated Macroporous Polypropylene Membranes for the Sustained Release of Drugs. *Nanomaterials* **2017**, *7*, 155. [CrossRef] [PubMed]
6. Sinn Aw, M.; Kurian, M.; Losic, D. Non-eroding drug-releasing implants with ordered nanoporous and nanotubular structures: Concepts for controlling drug release. *Biomater. Sci.* **2014**, *2*, 10–34. [CrossRef]
7. McInnes, S.J.; Irani, Y.; Williams, K.A.; Voelcker, N.H. Controlled drug delivery from composites of nanostructured porous silicon and poly (l-lactide) Research Article. *Nanomedicine* **2012**, *7*, 995–1016. [CrossRef] [PubMed]
8. Costa, P.; Sousa Lobo, J.M. Modeling and comparison of dissolution profiles. *Eur. J. Pharm. Sci.* **2001**, *13*, 123–133. [CrossRef]
9. Singhvi, G.; Singh, M. In Vitro drug release characterization models. *Int. J. Pharm. Stud. Res.* **2011**, *II*, 77–84.
10. Kikuchi, A.; Okano, T. Pulsatile drug release control using hydrogels. *Adv. Drug Deliv. Rev.* **2002**, *54*, 53–77. [CrossRef]
11. Lin, C.C.; Metters, A.T. Hydrogels in controlled release formulations: Network design and mathematical modeling. *Adv. Drug Deliv. Rev.* **2006**, *58*, 1379–1408. [CrossRef] [PubMed]
12. Carregal-Romero, S.; Guardia, P.; Yu, X.; Hartmann, R.; Pellegrino, T.; Parak, W.J. Magnetically triggered release of molecular cargo from iron oxide nanoparticle loaded microcapsules. *Nanoscale* **2015**, *7*, 570–576. [CrossRef] [PubMed]
13. Zhang, L.; Xia, J.; Zhao, Q.; Liu, L.; Zhang, Z. Functional graphene oxide as a nanocarrier for controlled loading and targeted delivery of mixed anticancer drugs. *Small* **2010**, *6*, 537–544. [CrossRef] [PubMed]
14. Alba, M.; Delalat, B.; Formentín, P.; Rogers, M.L.; Marsal, L.F.; Voelcker, N.H. Silica Nanopills for Targeted Anticancer Drug Delivery. *Small* **2015**, *11*, 4626–4631. [CrossRef] [PubMed]
15. Romero, V.; Vega, V.; García, J.; Prida, V.; Hernando, B.; Benavente, J. Effect of Porosity and Concentration Polarization on Electrolyte Diffusive Transport Parameters through Ceramic Membranes with Similar Nanopore Size. *Nanomaterials* **2014**, *4*, 700–711. [CrossRef] [PubMed]
16. Kumeria, T.; Santos, A.; Rahman, M.M.; Ferré-Borrull, J.; Marsal, L.F.; Losic, D. Advanced Structural Engineering of Nanoporous Photonic Structures: Tailoring Nanopore Architecture to Enhance Sensing Properties. *ACS Photonics* **2014**, *1*, 1298–1306. [CrossRef]
17. Santos, A.; Formentín, P.; Pallarès, J.; Ferré-Borrull, J.; Marsal, L.F. Structural engineering of nanoporous anodic alumina funnels with high aspect ratio. *J. Electroanal. Chem.* **2011**, *655*, 73–78. [CrossRef]
18. Santos, A.; Kumeria, T.; Wang, Y.; Losic, D. Insitu monitored engineering of inverted nanoporous anodic alumina funnels: On the precise generation of 3D optical nanostructures. *Nanoscale* **2014**, *6*, 9991–9999. [CrossRef] [PubMed]
19. Hanawa, T. Materials for metallic stents. *J. Artif. Organs* **2009**, *12*, 73–79. [CrossRef] [PubMed]
20. Swan, E.E.L.; Popat, K.C.; Grimes, C.A.; Desai, T.A. Fabrication and evaluation of nanoporous alumina membranes for osteoblast culture. *J. Biomed. Mater. Res. Part A* **2005**, *72*, 288–295. [CrossRef] [PubMed]
21. Osmanbeyoglu, H.U.; Hur, T.B.; Kim, H.K. Thin alumina nanoporous membranes for similar size biomolecule separation. *J. Membr. Sci.* **2009**, *343*, 1–6. [CrossRef]
22. Porta-i-Batalla, M.; Eckstein, C.; Xifré-Pérez, E.; Formentín, P.; Ferré-Borrull, J.; Marsal, L.F. Sustained, Controlled and Stimuli-Responsive Drug Release Systems Based on Nanoporous Anodic Alumina with Layer-by-Layer Polyelectrolyte. *Nanoscale Res. Lett.* **2016**, *11*, 372. [CrossRef] [PubMed]
23. Jeon, G.; Yang, S.Y.; Kim, J.K. Functional nanoporous membranes for drug delivery. *J. Mater. Chem.* **2012**, *22*, 14814. [CrossRef]

24. Xifre-Perez, E.; Guaita-Esteruelas, S.; Baranowska, M.; Pallares, J.; Masana, L.; Marsal, L.F. In Vitro Biocompatibility of Surface-Modified Porous Alumina Particles for HepG2 Tumor Cells: Toward Early Diagnosis and Targeted Treatment. *ACS Appl. Mater. Interfaces* **2015**, *7*, 18600–18608. [CrossRef] [PubMed]
25. Ribes, À.; Xifré-Pérez, E.; Aznar, E.; Sancenón, F.; Pardo, T.; Marsal, L.F.; Martínez-Máñez, R. Molecular gated nanoporous anodic alumina for the detection of cocaine. *Sci. Rep.* **2016**, *6*, 38649. [CrossRef] [PubMed]
26. Pla, L.; Xifre-Perez, E.; Ribes, A.; Aznar, E.; Marcos, M.D.; Marsal, L.F.; Martínez-Mañez, R.; Sancenon, F. A Mycoplasma Genomic DNA Probe using Gated Nanoporous Anodic Alumina. *Chempluschem* **2017**, *82*, 337–341. [CrossRef]
27. Vázquez, M.I.; Romero, V.; Vega, V.; García, J.; Prida, V.M.; Hernando, B.; Benavente, J. Morphological, Chemical Surface, and Diffusive Transport Characterizations of a Nanoporous Alumina Membrane. *Nanomaterials* **2015**, *5*, 2192–2202. [CrossRef] [PubMed]
28. Santos, A.; Ferré-Borrull, J.; Pallarès, J.; Marsal, L.F. Hierarchical nanoporous anodic alumina templates by asymmetric two-step anodization. *Phys. Status Solidi Appl. Mater. Sci.* **2011**, *208*, 668–674. [CrossRef]
29. Marsal, L.F.; Vojkuvka, L.; Formentin, P.; Pallarés, J.; Ferré-Borrull, J. Fabrication and optical characterization of nanoporous alumina films annealed at different temperatures. *Opt. Mater.* **2009**, *31*, 860–864. [CrossRef]
30. Santos, A.; Vojkuvka, L.; Alba, M.; Balderrama, V.S.; Ferre-Borrull, J.; Pallares, J.; Marsal, L.F. Understanding and morphology control of pore modulations in nanoporous anodic alumina by discontinuous anodization. *Phys. Status Solidi Appl. Mater. Sci.* **2012**, *209*, 2045–2048. [CrossRef]
31. Mardilovich, P.P.; Govyadinov, A.N.; Mukhurov, N.I.; Rzhevskii, A.M.; Paterson, R. New and Modified Anodic Alumina Membranes. 1. Thermotreatment of Anodic Alumina Membranes. *J. Memb. Sci.* **1995**, *98*, 131–142. [CrossRef]
32. Losic, D.; Santos, A. *Nanoporous Alumina: Fabrication, Structure, Properties and Applications*; Springer: Adelaide, Australia, 2015.
33. Han, H.; Park, S.J.; Jang, J.S.; Ryu, H.; Kim, K.J.; Baik, S.; Lee, W. In situ determination of the pore opening point during wet-chemical etching of the barrier layer of porous anodic aluminum oxide: Nonuniform Impurity Distribution in Anodic Oxide. *ACS Appl. Mater. Interfaces* **2013**, *5*, 3441–3448. [CrossRef] [PubMed]
34. Iskakov, R.M.; Kikuchi, A.; Okano, T. Time-programmed pulsatile release of dextran from calcium-alginate gel beads coated with carboxy-n-propylacrylamide copolymers. *J. Control. Release* **2002**, *80*, 57–68. [CrossRef]
35. Aukunuru, J.V.; Sunkara, G.; Ayalasomayajula, S.P.; Deruiter, J.; Clark, R.C.; Kompella, U.B. A biodegradable injectable implant sustains systemic and ocular delivery of an aldose reductase inhibitor and ameliorates biochemical changes in a galactose-fed rat model for diabetic complications. *Pharm. Res.* **2002**, *19*, 278–285. [CrossRef] [PubMed]
36. Horcajada, P.; Chalati, T.; Serre, C.; Gillet, B.; Sebrie, C.; Baati, T.; Eubank, J.F.; Heurtaux, D.; Clayette, P.; Kreuz, C.; et al. Porous metal–organic-framework nanoscale carriers as a potential platform for drug delivery and imaging. *Nat. Mater.* **2010**, *9*, 172–178. [CrossRef] [PubMed]
37. Vallet-Regí, M.; Balas, F.; Arcos, D. Mesoporous materials for drug delivery. *Angew. Chem.* **2007**, *46*, 7548–7558. [CrossRef] [PubMed]
38. Shoaib, M.H.; Tazeen, J.; Merchant, H.A.; Yousuf, R.I. Evaluation of Drug Release Kinetics From Ibuprofen Matrix Tablets Using Hpmc. *Pak. J. Pharm. Sci.* **2006**, *19*, 119–124. [PubMed]
39. Higuchi, T. Rate of release of medicaments from ointment bases containing drugs in suspension. *J. Pharm. Sci.* **1961**, *50*, 874–875. [CrossRef] [PubMed]
40. Higuchi, T. Mechanism of Sustained-Action Medication. Theoretical Analysis of Rate of Release of Solid Drugs Dispersed in Solid Matrices. *J. Pharm. Sci.* **1963**, *52*, 1145–1149. [CrossRef] [PubMed]
41. Brophy, M.R.; Deasy, P.B. Application of the Higuchi model for drug release from dispersed matrices to particles of general shape. *Int. J. Pharm.* **1987**, *37*, 41–47. [CrossRef]
42. Suvakanta, D.; Murthy, P.N.; Nath, L.; Prasanta, C. Kinetic Modeling on Drug Release from Controlled Drug Delivery Systems. *Pol. Pharm. Soc.* **2010**, *67*, 217–223.

nanomaterials

MDPI

Review

Review of Fabrication Methods, Physical Properties, and Applications of Nanostructured Copper Oxides Formed via Electrochemical Oxidation

Wojciech J. Stepniowski [1,2,*] and Wojciech Z. Misiolek [1]

[1]   Materials Science and Engineering Department & Loewy Institute, Lehigh University, 5 East Packer Ave., Bethlehem, PA 18015, USA; wzm2@lehigh.edu
[2]   Department of Advanced Materials and Technologies, Faculty of Advanced Technology and Chemistry, Military University of Technology, Urbanowicza 2 Str., 00-908 Warszawa, Poland
*   Correspondence: wos218@lehigh.edu or wojciech.stepniowski@wat.edu.pl; Tel.: +1-484-516-6544

Received: 31 March 2018; Accepted: 24 May 2018; Published: 29 May 2018

Abstract: Typically, anodic oxidation of metals results in the formation of hexagonally arranged nanoporous or nanotubular oxide, with a specific oxidation state of the transition metal. Recently, the majority of transition metals have been anodized; however, the formation of copper oxides by electrochemical oxidation is yet unexplored and offers numerous, unique properties and applications. Nanowires formed by copper electrochemical oxidation are crystalline and composed of cuprous (CuO) or cupric oxide (Cu_2O), bringing varied physical and chemical properties to the nanostructured morphology and different band gaps: 1.44 and 2.22 eV, respectively. According to its Pourbaix (potential-pH) diagram, the passivity of copper occurs at ambient and alkaline pH. In order to grow oxide nanostructures on copper, alkaline electrolytes like NaOH and KOH are used. To date, no systemic study has yet been reported on the influence of the operating conditions, such as the type of electrolyte, its temperature, and applied potential, on the morphology of the grown nanostructures. However, the numerous reports gathered in this paper will provide a certain view on the matter. After passivation, the formed nanostructures can be also post-treated. Post-treatments employ calcinations or chemical reactions, including the chemical reduction of the grown oxides. Nanostructures made of CuO or Cu_2O have a broad range of potential applications. On one hand, with the use of surface morphology, the wetting contact angle is tuned. On the other hand, the chemical composition (pure Cu_2O) and high surface area make such materials attractive for renewable energy harvesting, including water splitting. While compared to other fabrication techniques, self-organized anodization is a facile, easy to scale-up, time-efficient approach, providing high-aspect ratio one-dimensional (1D) nanostructures. Despite these advantages, there are still numerous challenges that have to be faced, including the strict control of the chemical composition and morphology of the grown nanostructures, their uniformity, and understanding the mechanism of their growth.

Keywords: anodization; copper oxides; nanostructures; passivation; nanowires; nanoneedles; band gap

---

1. Introduction

Nanostructured anodic oxides have attracted the attention of researchers due to their ease of fabrication and tailored ordered morphology on the nanometric scale. What is more, the resulting chemical and physical properties lead to numerous potential applications [1]. The most frequently studied anodic oxides are hexagonally arranged anodic aluminum oxide (AAO) [1] and nanoporous or nanotubular anodic titanium oxide (ATO) [2]. Intensive research on those two nanostructured materials has incited significant progress in: electrochemical and optical sensing [3,4], nanofabrication [5], photonic crystals [6], information optical coding [7], filtration and kidney dialysis [8], drug releasing platforms [9], biomaterials performance [10], renewable energy harvesting [11,12], the removal of greenhouse gases [13], magnetic materials [14], surface-enhanced Raman spectroscopy [15], plasmonic materials [16], structural color generation [17], tunable contact angle surfaces [18,19], and tunable band gap materials [20].

Currently, the majority of transition metals have been tested as substrates for anodizing. Researchers have obtained nanostructured anodic oxides as the result of the electrochemical oxidation of: W [21,22], Sn [23], Zr [24,25], Zn [26,27], Nb [28,29], Fe [30], and FeAl [20,31]. The majority of the nanostructures obtained by transition metals anodization are composed of oxide, where the metallic element is at one fixed oxidation state. Furthermore, almost all of the oxides are nanoporous or nanotubular. So far, the only reported exception is anodically grown ZnO: in this case, the grown oxide is made of nanowires [26,27].

Another promising metal for oxide nanostructures fabrication via self-organized anodization is copper. Copper forms two oxides, namely cuprous oxide Cu_2O and cupric oxide, CuO, as well as their mixtures in various phases such as copper-rich Cu_4O_3 [32]. A demand for the simple synthesis of copper oxides' nanostructures is a result of the electronic properties of Cu_2O, CuO, and Cu_4O_3. CuO is reported to be a p-type semiconductor with a band gap in the range from 1.2 to 2.16 eV, depending on the nature of the band gap (direct, or indirect), doping, morphology, and crystal size [32]. For CuO nanostructures, the smaller the size of the structure, the greater the band gap, observed as a blue shift in the spectrum. It is worth noting that Bohr's radius for CuO is ca. 6.6 nm, thus below this size a strong quantum confinement (QC) is observed [32] (nevertheless, QC is also observed above 6.6 nm, although the phenomenon is much weaker). Cu_2O is also a p-type semiconductor with a band gap over 2.1 eV [32]. However, also in this case, due to the QC, the band gap can be engineered and controlled during the manufacturing process. For example, a decrease of Cu_2O film thickness from 5.4 to 0.75 nm increases the band gap from 2.6 to 3.8 eV [33]. Additionally, according to Musselman et al., Cu_2O, due to the band gap value around 2.0 eV and theoretical maximum power conversion efficiency of approximately 20% (PCE), is suitable for applications in heterojunction solar cells [34].

There are numerous methods of Cu_2O and CuO nanostructures fabrication, employing diverse chemical methods including micelle-assisted precipitation, sol-gel methods, and high-temperature annealing in an oxidative atmosphere. The formation of a wide range of nanostructures such as nanospheres, nanoflowers, leaf-shaped nanocrystals, nanorings, nanoribbons, multi-pod nanocrystals, etc. has been already reported and broad reviews reporting the state-of-the-art have been published [35,36]. Furthermore, this variety of nanostructures with high surface-to-bulk atom ratios has triggered research on numerous applications of Cu_2O and CuO nanostructures, such as high-surface area electrode materials in batteries, photocatalysts in water purification systems, carbon dioxide reduction and water splitting, high-contact angle functional surfaces, gas sensors, infrared radiation sensors, etc. [35,36].

However, the anodization of copper, leading to the formation of nanostructures, has not been as intensively explored as other methods and was not even included in the numerous review reports focusing on the formation of cupric and cuprous oxide nanostructures. However, copper anodization may bring unexpected benefits in terms of morphology control, the formation of high-aspect ratio nanostructures, and their doping. Morphological features of nanostructured anodic oxides are tailored by operating conditions. For example, the pore diameter and intepore distance of anodic alumina and titania are linear functions of the applied voltage [1,2,20]. Additionally, electrochemical in situ doping of anodic oxides, due to the application of various additives in the electrolyte, allows one to dope the growing nanostructures [37]. Therefore, self-organized anodization seems to be a promising method in copper oxides formation, providing high-surface area nanostructures with a band gap tunable by operating conditions (size of the nanostructures) and chemical composition (in situ doping). Per analogiam to the anodization of Al and Ti [1–37], it can be expected that the morphology of the nanostructures formed by copper anodization can be tailored after the optimization of the operating conditions. Moreover, recent advances in Al and Ti anodization, as well as recent high-tech applications, may provide some inspiration for electrochemical copper oxidation.

2. Passivation of Copper

Pourbaix diagrams (potential vs. pH diagram) provide key information for understanding the electrochemical oxidation of metals. According to the Pourbaix diagram for copper, it is apparent that the most suitable electrolytes for copper anodization are the alkaline ones (solutions of bases like NaOH, KOH, but also salts with alkaline hydrolysis like carbonates and bicarbonates of potassium and sodium, not researched yet as potential copper anodizing electrolytes) (Figure 1) [38]. It is also noticeable that electrochemical oxidation may lead to the formation of oxides like Cu_2O (lower potentials) and CuO (greater potentials), as well as cupric hydroxide $Cu(OH)_2$ and water-soluble coordination anions with hydroxyl ligands, namely $Cu(OH)_3^-$ and $Cu(OH)_4^{2-}$. For anodizing, the minimal solubility of Cu_2O in water is at a pH range from 7.5 to 8.0 [38]. It is also worth noting that the as-obtained anodic oxides formed on copper are crystalline, while other anodic oxides, like titania or alumina, are amorphous. The formation of crystalline phases in anodic alumina or titania requires annealing after the anodization.

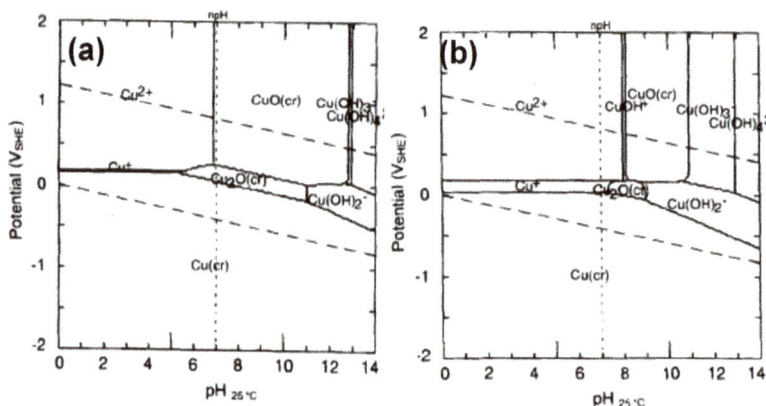

Figure 1. Pourbaix diagram for copper at 25 °C for total copper species concentration in a solution equal to 10^{-6} mol/L (a) and 10^{-8} mol/L (b). Reproduced with permission from [38]. Electrochemical Society, 1997.

Due to the formation of two copper oxides and the formation of soluble coordination ions, the mechanism of copper anodization is much more complex than in the case of other oxides, like AAO. Gennero de Chiavlo et al. analyzed the oxidation of copper only to the Cu^+ oxidation state and, already at this stage, the occurring phenomena are complex [39]. According to the chemical reactions, first copper oxidizes and forms metastable CuOH on the Cu surface (1):

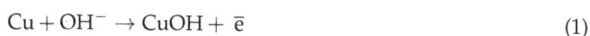

$$Cu + OH^- \rightarrow CuOH + \bar{e} \tag{1}$$

Next, cuprous hydroxide, CuOH, may react in two ways, either forming solid Cu_2O (2):

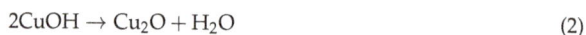

$$2CuOH \rightarrow Cu_2O + H_2O \tag{2}$$

Or bonding the hydroxyl group and forming water-soluble $Cu_2O_2H^-$ (3):

$$2CuOH + OH^- \rightarrow Cu_2O_2H^- + H_2O \tag{3}$$

In an alkaline environment, $Cu_2O_2H^-$ may also transform easily into $Cu_2O_2^{2-}$, when OH^- accepts a proton.

The cuprous oxide, under the influence of hydroxyl anions, may transform into the water-soluble $Cu_2O_2^{2-}$ (4):

$$Cu_2O + 2OH^- \rightarrow Cu_2O_2^{2-} + H_2O \tag{4}$$

Furthermore, the formed $Cu_2O_2^{2-}$, anions are metastable and disproportionate into solid, metallic Cu and water-soluble CuO_2^{2-}, which can be also considered as $[Cu(OH)_4]^{2-}$ (5):

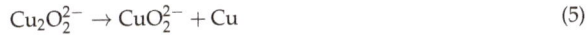

$$Cu_2O_2^{2-} \rightarrow CuO_2^{2-} + Cu \tag{5}$$

Thus, the re-deposited copper may undergo the entire cycle of reactions again, starting from Reaction (1). According to Ambrose et al., and their voltammetric study of Cu in KOH, a water-soluble Cu(I) species may also be formed directly from Cu, due to the formation of a coordination anion (6) [40]:

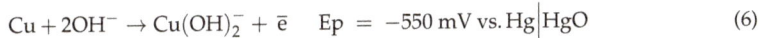

$$Cu + 2OH^- \rightarrow Cu(OH)_2^- + \bar{e} \quad Ep = -550\,mV\,vs.\,Hg|HgO \tag{6}$$

However, according to Reference [40], Cu_2O may also be electrochemically formed by the following Reaction (7):

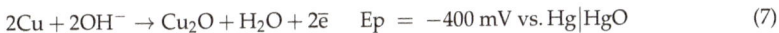

$$2Cu + 2OH^- \rightarrow Cu_2O + H_2O + 2\bar{e} \quad Ep = -400\,mV\,vs.\,Hg|HgO \tag{7}$$

The mechanism becomes more complex when the formation of Cu(II) from metallic Cu and Cu(I) is considered. Copper may oxidize directly to the soluble species like the coordination anion (8) [40]:

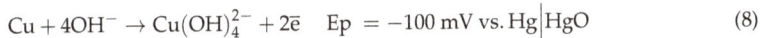

$$Cu + 4OH^- \rightarrow Cu(OH)_4^{2-} + 2\bar{e} \quad Ep = -100\,mV\,vs.\,Hg|HgO \tag{8}$$

However, the already-grown cuprous oxide, Cu_2O, may undergo further oxidation to the abovementioned soluble species (Equations (4) and (5)), or may form insoluble cupric hydroxide:

$$Cu_2O + 2OH^- + H_2O \rightarrow 2Cu(OH)_2 + 2\bar{e} \quad Ep = -100\,mV\,vs.\,Hg|HgO \tag{9}$$

Nevertheless, this $Cu(OH)_2$ deposit may form a soluble species like $[Cu(OH)_4]^{2-}$ in an alkaline environment [40]. According to the fundamental work by Ambrose et al. [40], at greater potentials Cu(II) may be directly formed by copper oxidation, according to the following reactions (10) and (11):

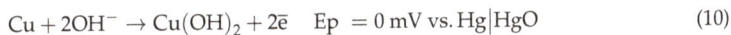

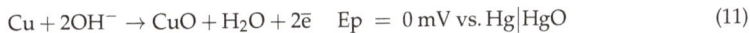

$$Cu + 2OH^- \rightarrow Cu(OH)_2 + 2\bar{e} \quad Ep = 0\,mV\,vs.\,Hg|HgO \tag{10}$$

$$Cu + 2OH^- \rightarrow CuO + H_2O + 2\bar{e} \quad Ep = 0\,mV\,vs.\,Hg|HgO \tag{11}$$

However, CuO and $Cu(OH)_2$ may also transform into soluble species, with copper at a greater oxidation state, namely Cu(III), when sufficiently high potentials are applied, as in reactions (12) and (13) [40]:

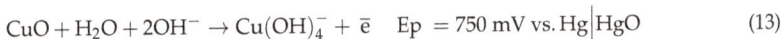

$$Cu(OH)_2 + 2OH^- \rightarrow Cu(OH)_4^- + \bar{e} \quad Ep = 750\,mV\,vs.\,Hg|HgO \tag{12}$$

$$CuO + H_2O + 2OH^- \rightarrow Cu(OH)_4^- + \bar{e} \quad Ep = 750\,mV\,vs.\,Hg|HgO \tag{13}$$

As can be deduced, from the abovementioned consideration, copper may form various chemical products at various oxidation states when it is electrochemically oxidized. This is in opposition to the anodization of the majority of transition metals. For example, anodizing aluminum provides only Al^{3+} species, namely Al_2O_3 and in some cases AlO(OH). Therefore, numerous copper anodization products, as well as the much more complex mechanism of growth, provide new areas to explore.

3. Strategies of Copper Anodization

The passivation of copper, due to the abovementioned complexity, triggers numerous topics and needs for fundamental research. The easiest way to passivate electrochemically metal, that has not yet been explored in terms of anodization, is the application of a potentiostat with a three-electrode system, with which potentiostatic experiments in the passivity field are conducted according to Pourbaix diagram. Stępniowski et al. reported such a study on Cu passivation in 1 M aqueous solution of KOH [41]. The voltammetric study showed without any doubt two distinct oxidation peaks, at ca. -450 and -150 mV vs. Ag|AgCl, responsible for metallic copper oxidation to Cu^+ and Cu^{2+}, respectively. Moreover, the morphology of the obtained oxides strongly depends on the applied potential: for low potentials, micron-sized cubes were formed, mainly made of cuprous oxide, while at -200 and -100 mV vs. Ag|AgCl, nanowires were grown that were composed of a mixture of cuprous and cupric oxide (Figure 2). Nevertheless, a photoluminescence study revealed the presence of CuO on the surfaces of all of the samples, shedding some light on the growth mechanism. This suggests, that oxidation first occurs from Cu° to Cu^+ and then, at the surface, Cu^+ oxidizes to Cu^{2+}. This is in line with the above-described considerations of anodic oxides' growth on copper [40]. It is noteworthy that the formed nanowires grow gathered in bundles. For example, those obtained at -200 mV (Figure 2C) grew in bundles 72 ± 14 nm thick, while individual nanowires were 24 ± 5 nm thick. Those formed at -100 mV were gathered into bundles 90 ± 23 nm thick, while the individual nanowires were 19 ± 7 nm thick [41]. According to Allam and Grimes, the morphology of the grown oxide can be diverse, but it can be controlled by the operating conditions, including various additives to the electrolyte [42]. They applied a two-electrode system, using KOH with various salts as additives, in order to check the influence of the additives on the morphology of the formed oxide. It was found that nanoneedles formed in a pure aqueous solution of KOH (pH = 11), while an addition of halogen salts such as NH_4F and NH_4Cl allowed the formation of micrometric crystals rather than nanostructures (Figure 3).

Furthermore, anodizing in ethylene glycol containing KOH and NH_4F also did not allow the achievement of nanostructures, like those anodized in an aqueous electrolyte. Leaf-like structures were obtained instead. Table 1 summarizes exemplary anodizing recipes in KOH-based electrolytes.

Table 1. Gathered experimental conditions for nanostructures formed via copper anodization in KOH-based solutions.

Chemical Composition of the Electrolyte	Experimental Conditions	Morphology and Chemical Composition of the Oxide	Remarks	Reference
0.25 M 0.5 M 1.0 M $K_2C_2O_4$ aq.	Cyclic voltammetry: 50 mV/s Range: −1.0 to 0.8 V RT [1]	CuO/CuO_x Microporous structure	Anodized surface was used as an electrochemical sensor of glucose; glucose was determined at concentrations as low as 4 mM in human blood serum using anodized copper	Satheesh Babu 2010 [43]
Aqueous KOH	4-6 V RT	Nanoneedles	No nanostructures were formed in aqueous KOH at pH < 10	Allam 2011 [42]
Aqueous KOH	10 V, pH = 11, 11.5, 12	Nanorods	At a pH below 10, Cu was dissolved; light blue precipitate was formed on the samples; surface nanostructuring enhanced the photoelectrochemical response	Shooshtari 2016 [44]
2 M KOH	1.5 mA/cm², varied duration	$Cu(OH)_2$ nanoneedles	Cu mesh was anodized in order to change the contact angle (CA); CA was pH-responsive: the lower the pH, the greater the CA, up to 153°	Cheng 2012 [45]
0.5-4.0 M KOH	0.5–4.0 mA/cm², 5–25 °C, 25 min	CuO nanoneedles 170 ± 40 nm in diameter 7–10 μm long	Fluoroalkyl-silane (FAS-17) was chemically bonded to CuO nanoneedles, increasing the contact angle up to 169°	Xiao 2015 [46]
2, 2.5, 3, 3.5 M KOH	1.5 mA/cm², 2, 15, 28 °C	$Cu(OH)_2$ and CuO nanoneedles 500-550 nm in diameter	$Cu(OH)_2$ were turned into CuO nanoneedles using heat treatment (150 °C at 3 h + 200 °C at 3 h); in 3 M KOH at 28 °C nanotubes were formed (80-500 nm diameter, 10 μm length)	Wu 2005 [47]
1 M KOH	−0.4, −0.3, −0.2, −0.1 V vs. Ag | AgCl, RT, 1 h	Morphology depends on the potential; cubes and nanowires	Formed nanowires were mixtures of Cu_2O and CuO; nanowires were obtained for −0.2 (24 nm in diameter) and −0.1 V (19 nm in diameter), while for −0.4 and −0.3 V micro-cubes were formed	Stepniowski 2017 [41]
0.15 M KOH + 0.1 M NH_4Cl	6 V, 300 s, RT	Cubes and dendrites	-	Allam 2011 [42]
0.15 M KOH + 0.1 M NH_4F	6 V, 300 s, RT	Cu_2O Micro-balls made of whiskers forming nanopores	[3] XPS and [4] GAXRD proved that the structures were made of Cu_2O	Allam 2011 [42]
0.15 M KOH + 0.1 M NH_4F + 3% H_2O in EG [2]	30 V, 300 s, RT	160-nm thick leaf-like architectures	When KOH concentration was increased to 0.2 M, the leaf-like structures were ca. 500 nm thick; at voltages below 30 V, no structured film was formed	Allam 2011 [42]
0.75 wt % KOH + 3 wt % H_2O + 0.20–0.35 wt % NaF in EG	10-30 V, 10 min	Cu_2O film	Cu_2O film was formed by anodization; further annealing (250–450 °C, 60 min) allowed the growth of CuO nanowires, improving the photoelectrochemical performance	Wang 2013 [48]
0.1–0.5 M KOH + 0–0.1 wt % NH_4F + 1 vol % H_2O in EG	5–20 V, 5 °C	Nanoporous film (6–15 nm pore diameter)	Nanoporous oxide was formed, composed of a mixture of the following species: Cu_2O, CuO, $Cu(OH)_2$, and CuF_2	Oyarzún Jerez 2017 [49]

[1] RT—room temperature; [2] EG—ethylene glycol; [3] XPS—X-ray photoelectron spectroscopy, [4] GAXRD—glancing angle X-ray diffraction.

Generally, anodization in aqueous KOH electrolytes results in the formation of nanoneedles [42,45–47], nanowires [41], and nanorods [44]. Specifically, in copper anodization, nanoneedles are as long as nanowires but their diameter decreases close to their top, while nanowires have quite a uniform diameter along their length (nanorods are like nanowires, but with a much smaller aspect ratio). A wide range of the structures' diameter has been observed; while Stepniowski et al. reported the smallest nanowire diameter equal 19 nm [41], Xiao et al. [46] reported the formation of nanoneedles with a diameter of 170 nm (up to 10 μm long) and the formation of nanoneedles 500–550 nm thick was reported by Wu et al. [47]. Thus, this shows that the anodization of copper provides the possibility to tune the morphology of the grown oxides in a wide range, making this technique competitive in comparison to others. Numerous other techniques have limitations in morphology control, while anodization allows to one obtain high-aspect ratio one-dimensional (1D) nanostructures within a wide range of diameter.

Figure 2. Top-view FE-SEM images of the surface morphology of the oxides formed via copper passivation in 1.0 M KOH at −400 (**A**); −300 (**B**); −200 (**C**); and −100 mV (**D**). Reproduced with permission from [41]. Elsevier, 2017.

Figure 3. Top-view FE-SEM images of effects of Cu anodization in: (**a**) 0.15 M KOH + 0.1 M NH$_4$Cl at 6 V for 300 s; (**b**) 0.2 M KOH + 0.1 M NH$_4$F at 6 V for 300 s; and (**c**) aqueous solution of KOH (pH = 11) at 10 V. Reproduced with permission from [42]. Elsevier, 2011.

Additives, compounds added to the electrolytes, allow the modification of the morphology, as detailed in the abovementioned publication [42]. It is worth noting that recently a manufactured nanoporous material made of Cu$_2$O, CuO, Cu(OH)$_2$, and CuF$_2$ was obtained with a pore diameter ranging from 6 to 15 nm [49]. Typically, the majority of the anodic oxides formed on copper are made of nanowires, nanoneedles, or nanorods. Nanoporous morphology is typical for oxides grown on other metals. Moreover, the pores formed via copper anodization have smaller diameters than the majority obtained by aluminum anodization. However, the ordering of the pores formed on copper is much poorer, compared to highly-ordered AAO. Nevertheless, the formation of porous oxide on copper confirms the vast variety of morphologies possible to obtain by anodization.

It is also worth noting that, analogous to aluminum anodization, and in the case of copper anodization, both potentiostatic and galvanostatic approaches are used. At a constant voltage (potentiostatic approach), the voltage resulting in the formation of anodic oxides in KOH-based solutions ranges from 4 to 30 V in two-electrode systems. When the galvanostatic approach is used, the current density ranges from 0.5 to 4.0 mA/cm^2 (Table 1). Due to the alkaline hydrolysis, potassium oxalate was also successfully applied in copper anodization [43], producing a microporous mixture of CuO and CuO$_x$ due to the appropriate pH of the applied electrolyte (compare to the Pourbaix diagram, Figure 1). Thus, this shows there is still much to explore in the anodization of copper. THe Pourbaix diagram shows numerous opportunities for copper anodization at a more ambient pH. Carbonates and bicarbonates of alkali metals would be ideal for further fundamental research in this field.

Wu et al. recently reported the behavior of copper in 0.1 M NaOH during cyclic voltammetry scans. According to their findings, copper in NaOH oxidizes in a few stages: at low potentials, Cu$_2$O is first formed, then Cu$_2$O at the surface is oxidized to CuO, and finally, Cu(OH)$_2$ forms the outer layer of the oxide structures [50]. These findings are analogous to the recent results for Cu oxidation in 1 M KOH [41] and they are also in line with the Pourbaix diagram, where at lower potentials copper oxidized to Cu$_2$O and then, at greater potentials, oxidized to CuO (Figure 1). Anodization in NaOH-based solutions also allows the formation of various nanostructures. The most desired nanostructures are nanoneedles and nanowires, due to their high surface area, like those formed in 1 M NaOH (Figure 4). Table 2 summarizes the recipes for copper anodization in NaOH-based solutions.

Figure 4. FE-SEM images of nanostructures grown on copper in 1.0 M NaOH at −200 mV vs. Ag|AgCl for 10 min at room temperature (RT). Images taken at different magnifications (**A–C**). Unpublished research by Stępniowski et al.

Table 2. Gathered experimental conditions for nanostructures formed via copper anodization in NaOH-based solutions.

Chemical Composition of the Electrolyte	Experimental Conditions	Morphology and Chemical Composition of the Oxide	Remarks	Reference
0.1 M NaOH	−400 mV, 1 h	Nanoparticles	Mechanism of oxide growth was studied	Caballero-Briones 2010 [51]
0.1 M NaOH	10 mV/s voltammetric scan from −1.2 to 0.8 V RT	Cu needle was anodized and coated by oxide-hydroxide film	Mechanism of Cu electrochemical oxidation was investigated	Wu 2013 [50]
1 M NaOH	0.06 mA/cm², 5 min, 25 °C	Cu(OH)₂ nanowires	Nanowires surface was modified by the chemical bonding of 1H,1H,2H,2H-Perfluorodecyltriethoxysilane (FAS-17) in order to increase the wetting contact angle to 154°; as-obtained nanowires were made of Cu(OH)₂, but further annealing enabled the transofrmation of the hydroxide into CuO	Jiang 2015 [52]
1 M NaOH	Cyclic voltammetry from −1.6 V to 0.4 V RT	CuO dendrite crystals grown on Cu₂O nanoparticles	Cyclic voltammetric study of Cu in alkaline solution	Wan 2013 [53]
3 M NaOH	1.5, 3.0, 5.0 mA/cm², 30 min	Cu₂O, Cu₂O/Cu(OH)₂, Cu(OH)₂ nanowires	Cu was electrodeposited on ITO (Indium Tin Oxide) and subsequently anodized; obtained nanostructures enhanced photocatalytic water splitting; the best results were achieved for nanowires made of both Cu₂O and CuO	Zhang 2012 [54]
0.15 M NaOH	pH = 12.8–13.0 −430 mV	Cu₂O nanoparticles	Cu₂O behaved like a p-type semiconductor	Caballero-Briones 2009 [55]
1 M NaOH + 2.5 M NaCl + 0.5 g/L EG	0.5–2.5 A/dm², 55–70 °C, 30 min	30-nm thick nanosheets made of CuO and Cu₂O	Formed nanostructures are mixtures of CuO and Cu₂O	Shu 2017 [56]
10 wt % NaOH + 5 wt % NaClO₂	0.75 V, 60 °C, 15 min	CuO films with traces of Cu₃O₂, decomposing to CuO and Cu₂O	After anodizing, samples were immersed in KMnO₄; consequently, solar light absorption reached up to 96%	Arurault 2007 [57]
0.2 M NH₄Cl; pH was adjusted to 8 with NaOH	5 mA/cm², RT, 20 min	Cu(OH)₂ film	In order to transform Cu(OH)₂ into Cu₂O, three post-treatments were applied: hydrolysis in H₂O₂, reduction in H₂ at 280 °C, redox with glucose; the H₂O₂ post-treated sample had the best efficiency in oxygen generation	Zhang 2015 [58]

According to Table 2, it is apparent that various approaches were tested to grow anodic oxides on Cu in NaOH-based electrolytes. Here also the three-electrode approach using cyclic voltammetry [50,53] or constant potential [51,55,57] was applied. In a two-electrode approach, galvanostatic anodizing was applied with current densities in the range of 0.06 mA/cm^2 [52] up to 5 mA/cm^2 [54,58]. The concentration of NaOH in the applied electrolyte was used in the range from 0.1 [50,51] to 3.0 M [54], but also in this case, various additives were used to grow the nanostructured oxides, namely: NaCl with ethylene glycol [56], NaClO$_2$ in order to enhance oxidation [57], and NH$_4$Cl [58]. The application of sodium hydroxide-based electrolytes allowed the achievement of oxides with various morphologies, such as nanoparticles [51,53,55], nanowires [52,54], dendrites [53], nanoneedles [50], and nanosheets [56]. Also in this case, copper anodization offered a wide range of nanoneedle diameters and a high aspect ratio, overcoming the limitations of other techniques (Table 2).

In further applications, especially in physical ones such as photovoltaics, chemical cleanness is crucial. Unfortunately, in the majority of the research, detailed chemical composition analyses, mainly X-ray photoelectron spectroscopy (XPS), have revealed the simultaneous presence of Cu$_2$O, CuO, or even Cu(OH)$_2$ in the grown nanostructures [56–58]. For energy harvesting applications, such as PV (photovoltaics) or photocatalytic water decomposition, Cu$_2$O is the most demanded among the copper species and its nanostructured form provides numerous surpluses thanks to its high surface area. To face the challenge of chemical composition homogeneity, Zhang et al. reported three methods of transforming Cu(OH)$_2$ nanostructures into Cu$_2$O [58]. One of them was hydrolysis with hydrogen peroxide (14):

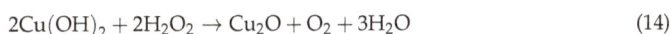

$$2Cu(OH)_2 + 2H_2O_2 \rightarrow Cu_2O + O_2 + 3H_2O \qquad (14)$$

Another involved the annealing of the anodized samples in hydrogen at 280 °C (15):

$$2Cu(OH)_2 + 2H_2 \rightarrow Cu_2O + 3H_2O \qquad (15)$$

And the third involved a reaction with glucose (used in electrochemical glucose sensing) that allowed the reduction of Cu(OH)$_2$ to Cu$_2$O (16):

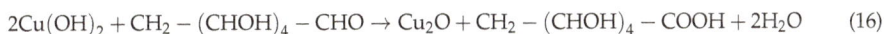

$$2Cu(OH)_2 + CH_2 - (CHOH)_4 - CHO \rightarrow Cu_2O + CH_2 - (CHOH)_4 - COOH + 2H_2O \qquad (16)$$

The obtained cuprous oxide had improved ability in photocatalytic oxygen generation from water, although the best results (the most efficient oxygen production) were achieved with the samples treated with hydrogen peroxide. This means that the reduction of copper from Cu(II) to Cu(I) was the most efficient. Therefore, the nanostructures obtained by Cu anodization can be chemically reduced, forming high-surface area material made of Cu$_2$O. Such numerous options of cuprous oxide as well as cupric oxide nanostructuring allow researchers to apply these nanostructures in various devices.

4. Properties and Applications of Anodic Nanostructures Grown on Copper

The high surface area and chemical composition that could be tailored via anodization encouraged scientists to apply these materials in areas such as tunable contact angle surfaces, sensing, and renewable energy harvesting.

As mentioned above, the copper species formed via anodization are at various oxidation states. Also, chemical post-treatments can be applied to oxidize or reduce the formed nanostructures. Thus, they can form redox couples with various chemical compounds, providing reactions that can be electrochemically sensed. One such reaction is glucose oxidation to gluconic or glucuronic acid (see Equation (16)). However, the C–C bonds in glucose may undergo dissociation during the electrochemical oxidation and form shorter products like formats, or even carbonates. Simultaneously, active surface Cu(III) species are reduced to Cu(II) [43]. Satheesh Babu and Ramachandran anodized Cu in potassium oxalate in order to obtain an electrochemical glucose sensor [43]. This allowed the formation of CuO with a developed surface area, which is desired in sensing applications. Applied potassium oxalate solution as the electrolyte allowed the incorporation of oxalates into the CuO, in the form of Cu(II) oxalate. This in situ doping approach has already been intensively researched for other anodic oxides, like alumina [37]. However, it was shown in this work [43] that the in situ doping of CuO is also possible and beneficial for sensing applications. The electrochemical sensor made of Cu coated with CuO and copper oxalate was found to have a sensitivity as high as 1.89 mA·mM^{-1}·cm^{-2} with a detection limit equal to 50 nM. Additionally, the sensor was found to work in the presence of other interfering compounds like ascorbic acid and uric acid, as well as in samples of human blood serum, confirming its outstanding selectivity towards glucose.

Due to the developed surface area and diverse morphologies of the formed anodic films, the modification of the wetting of the surface can be provided by copper anodization. There are two main approaches describing the behavior of liquids on surfaces with nanostructures on the top (and mixed approaches as well): Wenzel's approach, in which the nanostructures are deeply penetrated by the liquid, and Cassie-Baxter's method, in which the liquid does not penetrate the nanostructures and air remains trapped inside them. According to Jiang et al., the nanostructures grown via copper anodization are in the Cassie-Baxter state [52]. They anodized copper in 1 M NaOH (Table 2) and obtained Cu(OH)$_2$ nanowires that were hydrophilic (wetting contact angle was 4.5° and 0° for water and CH$_2$I$_2$, respectively). Due to the provided modification of the surface with the chemical bonding of FAS-17 (2H,2H-Perfluorodecyltriethoxysilane), the contact angle increased significantly to 154° and 133° for water and CH$_2$I$_2$, respectively [52]. FAS-17-modified CuO nanostructures were also used to research anti-corrosion applications of the as-formed hydrophobic coating. Xiao et al. reported the formation of CuO nanoneedles in KOH that were 7–10 μm long and ca. 170 nm in diameter (Table 1) (Figure 5A,B) [46]. Post-treatment with FAS-17 allowed the increase of the wetting water contact angle up to 169°, providing superhydrophobicity (Figure 5C,D). Furthermore, the limited contact between the surface and surrounding liquid environment (Cassie-Baxter state) hinders galvanic coupling with the material underneath, consequently improving the corrosion performance. After Cu anodization for 40 min in 2 M KOH at 2 mA·cm^{-2} at 15 °C, the corrosion potential increased from −254 to −212 mV, while the corrosion current density recorded in 3.5% NaCl decreased from 19.58 μA·cm^{-2} to 9.11 μA·cm^{-2} [46]. Further surface chemical modification by FAS-17 bonding increased the hydrophobicity and consequently increased the corrosion potential to −124 mV and decreased corrosion current to 0.66 μA·cm^{-2}. After one week of immersion in 3.5% NaCl, the corrosion performance of the anodized and chemically modified copper remained satisfactory: the corrosion current density was then 0.99 μA·cm^{-2} and the corrosion potential was −134 mV (Figure 5E,F). Thus, morphologically and chemically generated hydrophobicity could hinder the charge transfer at the electrolyte-sample interface, as confirmed by electrochemical impedance spectroscopy (EIS)-derived charge transfer resistance, which was 2.47 kΩ·cm^2 for anodized and modified samples vs. 0.25 kΩ·cm^2 for pristine Cu and 0.44 kΩ·cm^2 for the only anodized surface.

Another application, using the superhydrophobic effect admitted to the morphology of anodic oxides grown on copper, employs a pH-responsive water permeation mesh [45]. Cu(OH)$_2$ nanoneedles formed in KOH (see Table 1 and Figure 6A–C) were coated with gold and then thiols, HS(CH$_2$)$_9$CH$_3$ and HS(CH$_2$)$_{10}$COOH, were chemically bonded to the surface. At a low pH, such a nanostructured mesh had a contact angle equal 153°, which did not allow water to pass through the mesh (Figure 6D). However, increasing pH to 12 significantly decreased the contact angle (ca. 8°) and consequently the mesh was permeable to water [43]. What is even more interesting, the pH-response time was about 3 s. Moreover, this system was reversible and switching the pH allowed one to switch contact angle and, consequently, the permeability. It is also important to note that the starting material had a complex geometry in the micro scale (mesh) and it was successfully anodized. Therefore, this paper shows that samples with sophisticated morphology can be anodized, providing cuprous and cupric oxide nanostructures on the surface. Typically, when Al or Ti are anodized, the starting material samples are plates or rods. Very often, due to the complex geometry of the metallic substrate, anodization is impossible because the anodic dissolution of metals occurs at the exposed edges of the sample. Reporting the anodization of Cu mesh enables numerous opportunities in other applications, where already the starting material must have complex geometry, but formation of surface nanostructures would be beneficial and in-demand.

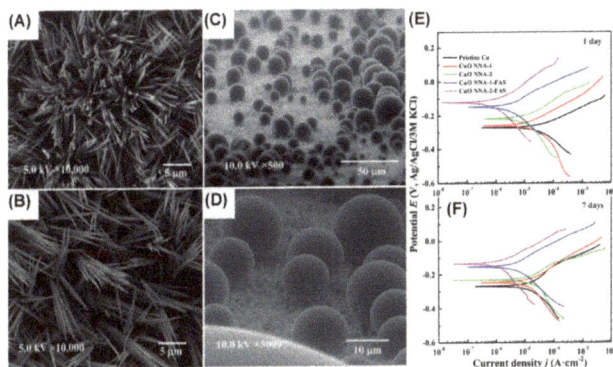

Figure 5. Top-view FE-SEM images of CuO nanoneedles formed in 2 M KOH at 15 °C, 2 mA·cm^{-2} for 25 (**A**) and 40 min (**B**); behavior of 3.5% NaCl solution on their surface after the chemical bonding of FAS-17 (**C,D**) and their corrosion performance after 1 (**E**) and seven days of immersion in 3.5% NaCl (**F**). CuO NAA-1 and CuO NAA-2 indicate anodization for 25 and 40 min, while FAS denotes the subsequent modification with FAS-17. Reproduced with permission from [46]. Royal Society of Chemistry, 2015.

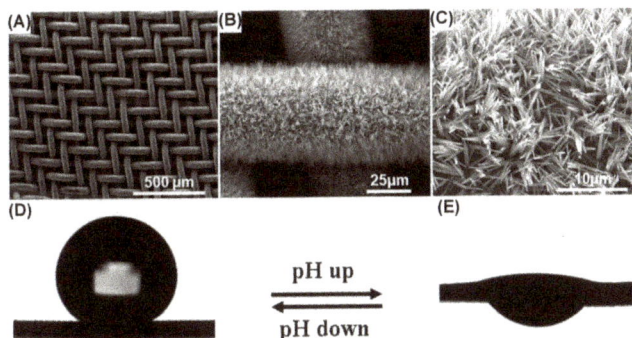

Figure 6. SEM images of Cu mesh subjected to anodization in 2 M KOH at 1.5 mA·cm^{-2} (**A–C**) and water permeation performance after anodization, Au sputtering, and functionalization with thiols (**E,F**); water droplet size was 4 μL (**D,E**) and pH was 2 (**D**) and 12 (**E**). Reproduced with permission from [45]. American Chemical Society, 2012.

One of the applications in which cupric oxide nanostructures formed via anodization may bring significant input is renewable energy harvesting [59]. Cu$_2$O (a p-type semiconductor) coupled with ZnO (an n-type semiconductor) can be used as an efficient hetero-junction photovoltaic cell. However, so far only around a 6% power conversion efficiency (PCE) of Cu$_2$O-based photovoltaic cells has been reported, versus 23% of the theoretical PCE value [59]. Numerous nanostructuring approaches, including the above-discussed anodization, may bring progress in the PCE. Due to the electronic structure of Cu$_2$O and CuO, nanostructures obtained by copper anodization will find applications in renewable fuel generation, namely in photoelectrochemical (PEC) water splitting. The semiconducting photocathode has a suitable band gap for water splitting reactions, which is within the band gap of Cu_2O and CuO.

It is important to note that the band gap of CuO is smaller than that of Cu_2O; moreover, both the valence and conduction bands of CuO are lower than those of Cu_2O. Briefly, after the excitation of the photocathode made of Cu_2O-CuO film, when the excited electron will exceed the band gap of Cu_2O, it can reduce its energy down to the conduction band of CuO and by further reduction cause the decomposition of water, generating gaseous hydrogen and oxygen. Nevertheless, one should be aware that corrosive reactions, like Cu_2O or CuO reduction, may occur as well, affecting the performance of the photocatalyst over time. Zhang et al. reported an extensive study on copper anodization for photoelectrochemical water splitting [56]. They investigated a few strategies of copper anodization, employing various current densities and post-treatments such as post-annealing, causing calcinations and improving the crystallinity of the formed nanoneedles (Table 2). The reported research confirmed that the mixed oxide nanoneedles, consisting of a Cu_2O-CuO system, have the best performance towards PEC water splitting. The greatest photocurrents were recorded for $Cu/Cu_2O/CuO$ and $Cu/Cu_2O/Cu(OH)_2$, namely -1.54 and -1.28 mA/cm^2, respectively, at no external voltage (0 vs. NHE — Normal Hydrogen Electrode). Nevertheless, due to the mentioned corrosion of cupric and cuprous oxide, their stability dropped over time. $Cu/Cu_2O/CuO$ and $Cu/Cu_2O/Cu(OH)_2$ preserved after 20 min represented 74.4% and 85.8% of their performance, respectively. Pure Cu_2O film, after 20 min of performance, had only 30.1% of its primal conversion efficiency (simultaneously harvesting a much lower current density: -0.65 mA/cm^2).

Photocatalytic oxygen generation is another method of renewable energy harvesting, enabling the storage of energy. In order to generate oxygen from water, photo-generated holes have to recombine on the surface of the catalyst, according to Equation (17) [58]:

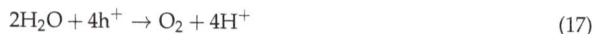

$$2H_2O + 4h^+ \rightarrow O_2 + 4H^+ \tag{17}$$

This reaction lays in the band gap of Cu_2O, thus nanostructures formed by Cu anodization are also suitable for this application. However, in contrast to PEC water splitting, in this case the presence of CuO would affect the yield of the reaction. Thus, the authors [58] worked out three various approaches to reduce Cu(II) to Cu(I), mentioned in the previous section (Equations (14)–(16)). According to the authors, the oxygen generation reaction occured efficiently due to the contact of metallic Cu with Cu_2O, as it allowed the rapid acceptance of electrons generated by metal, with a simultaneous rapid conversion of holes on the catalysts' surface. Thus, the formation of charge imbalance was hindered due to this junction. Among three post-treatment methods, hydrolysis with hydrogen peroxide obtained the greatest oxygen production yield, providing 233.27 µmol from 1 mg of catalyst in 8 h of its performance, while methods obtained results below 185 µmol of O_2 [58].

On the other hand, the post-treatment of the grown anodic oxides with $KMnO_4$ was reported by Arurault et al. [57]. This enabled the oxidization of the grown anodic film to CuO and Cu_3O_2. The highly-developed surface area combined with the chemical composition increased solar light absorption. Thus, it was demonstrated that anodic oxides grown on copper are suitable for solar cell applications.

A brief review of anodic oxides grown on copper with their applications is presented in Table 3. It leads to the conclusion that the major applications of anodic oxides grown on copper result from their chemical composition and nanostructured morphology. However, most applications are reported for anodic oxides in the form of nanoneedles. This form of nanostructure provides high-surface area for photocatalytic reactions high wetting contact angle (after functionalization), and high sensitivity.

Table 3. Gathered information about applications of nanostructures formed via copper anodization.

Application	Role of the Anodically Grown Nanostructures	Remarks	Reference
Glucose sensor	High surface area Cu(II) active sites were reducted to Cu(II) by glucose	Glucose was determined at concentrations as low as 4 mM in human blood serum	[43]
pH-responsive water permeation mesh	High surface area	Wetting contact angle was pH-switchable and for a lower pH it reached up to 153°	[45]
High contact angle surface for corrosion protection	Highly-developed nanostructured surface area	Fluoroalkyl-silane (FAS-17) was chemically bonded to CuO nanoneedles, increasing the contact angle up to 169°; the corrosion performance was significantly improved	[46]
High contact angle surface	Highly-developed nanostructured surface area	Nanowires' surface was modified by the chemical bonding of 1H,1H,2H,2H-Perfluorodecyltriethoxysilane (FAS-17) in order to increase the wetting contact angle to 154°	[52]
Photoelectrochemical chemical water splitting	Highly-developed surface area and chemical composition (CuO-Cu$_2$O)	Nanostructures enhanced photocatalytic water splitting; the best results were achieved for nanowires made of both Cu$_2$O and CuO	[54]
Photochemical oxygen generation	High surface area and chemical composition (Cu$_2$O)	Post-treatment of obtained nanostructures was conducted	[58]
Solar light absorption	Chemical composition (CuO with Cu$_3$O$_2$)	Post-treatment in KMnO$_4$ was conducted	[57]

5. Anodization versus Other Nanostructured Copper Oxide Fabrication Methods

According to the abovementioned applications, 1D cuprous and cupric oxide nanostructures may bring valuable contributions to many fields. Thus, a variety of the Cu_2O and CuO synthesis techniques are being developed by researchers. The literature study shows that the self-organized anodization of copper may be attractive for few reasons (Table 4):

- It employs inexpensive equipment;
- It is time-efficient;
- It is an easy technique to scale-up, as it employs self-organization (no template required);
- It does not require numerous steps, only electropolishing and one step of the electrochemical oxidation;
- It allows for the control of geometry and provides 1D nanostructures with a high aspect ratio.

On the other hand, there are numerous competitive chemical techniques, like precipitation, sol-gel, or hydrothermal synthesis, but 1D nanostructures obtained with those methods have much smaller aspect ratios. Moreover, the chemical syntheses are multistep processes. Additionally, some of the steps take tens of hours to complete. Template-assisted techniques that also employ atomic layer deposition (ALD) may provide high aspect ratio 1D nanostructures made of cuprous and cupric oxides; however, these methods are also multistep (the formation of template also must be taken into account) and employ expensive tools (ALD).

Therefore, the self-organized anodization of copper, resulting in the formation of 1D nanostructures, may nowadays attract much more attention due to the offered solutions and potential applications of the nanostructures.

Table 4. Drawbacks of currently applied copper oxides nanostructuring methods vs. solutions offered by anodizing.

Fabrication Method	Drawbacks of the Method	Solution offered by Anodizing	Reference
Hydrothermal synthesis	Small aspect ratio of one-dimensioanl (1D) nanostructures; requires a few steps of synthesis	High aspect ratio of 1D nanostructures can be easily achieved by lengthening the time of anodization; facile, easy-to-scale-up two-step synthesis (electropolishing + anodizing)	[35,36]
Atomic layer deposition	Requires templates and expensive equipment to grow 1D nanostructures	Method based on easy to scale-up self-organization; inexpensive, electrochemical method	[35,36]
Solution-based chemical precipitation	Small aspect ratio of 1D nanostructures	Relatively high aspect ratio (see Tables 1 and 2)	[35,36]
Template techniques (i.e., Anodic Aluminum Oxide, AAO and subsequent deposition)	Numerous steps of synthesis (formation of template, deposition, removal of the template)	An easy-to-conduct, two-step synthesis (electropolishing + anodizing)	[59]
Sol-gel techniques	Multistep process, time-consuming method	Anodization can be minutes long, in order to achieve a surface covered by oxide nanoneedles	[60]

6. Conclusions and Challenges

The self-organized anodization of copper allows one to obtain anodic oxides with various morphologies, including nanoneedles, nanowires, or even nanoporous structures. Furthermore, the chemical composition of the formed nanostructures can be varied using appropriate experimental conditions in anodizing and various post-treatment procedures. The literature reveals that nanostructured oxides grown on copper can contribute to glucose sensing in the presence of interfering compounds and in human blood serum. Furthermore, the nanoneedles grown by copper anodization are capable of forming superhydrophobic surfaces, including smart, pH-responsive permeation systems. Copper anodization also contributes to renewable energy harvesting research: it enables the photocatalytic decomposition of water to hydrogen and oxygen. When the photocatalyst is made of a Cu_2O-CuO mixture, it can generate oxygen from water.

The current state-of-the art review also revealed numerous challenges in copper anodization:

1. There are challenges in the formation of anodic copper oxides: generally, only KOH- and NaOH-based electrolytes were used for anodizing. The Pourbaix diagram revealed many more opportunities. Thus, electrolytes in a pH range from ca. 7 to 11 could provide different morphologies and more uniform chemical compositions. Potassium oxalate, used as an anodizing electrolyte, shows that there is still much to explore in this field.

2. A full systemic study of the influence of the operating conditions (type of the electrolyte, voltage, time, temperature) on the morphology of the grown nanostructures has not yet been reported, though this would be beneficial for all researchers working in the field of applications of copper and cuprous oxide nanostructures. In this field, the formation of nanostructures via anodization with a diameter smaller than ca. 7 nm, in order to observe a significant QC effect, would be also challenging.

3. There is also a demand to quantify the chemical composition of the grown nanostructures versus the operating conditions of anodization. It would be beneficial from both a fundamental point of view (understanding the mechanism of growth with hard, experimental data) and for applications in which chemical purity is crucial, such as photovoltaics.

4. In contrast to all anodic oxides, nanostructures grown by copper electrochemical oxidation are crystalline. From a fundamental point of view, investigating the crystal orientation would be important, so as to determine whether the anodization of a planar sample, in a constant electric field, induces the growth of oriented, crystalline nanowires and nanoneedles.

In summary, anodization, a relatively novel approach in copper oxides nanoengineering, allows one to obtain a variety of morphologies, thus contributing to state-of-the-art emerging applications.

Author Contributions: W.J.S. is responsible for the literature review and presenting the information in the tables. Both authors participated in the manuscript preparation; however, majority of the writing was done by W.J.S. and most of the editing was done by W.Z.M.

Acknowledgments: Wojciech J. Stępniowski cordially acknowledges financial support from the Kosciuszko Foundation in New York, NY, USA. Both authors would like to express their gratitude to the Loewy Family Foundation for the financial support through endowment at Lehigh University.

Conflicts of Interest: The authors declare no conflict of interest.

References

1. Jani, A.M.M.; Losic, D.; Voelcker, N.H. Nanoporous anodic aluminium oxide: Advances in surface engineering and emerging applications. *Prog. Mater. Sci.* **2013**, *58*, 636–704. [CrossRef]

2. Kowalski, D.; Kim, D.; Schmuki, P. TiO$_2$ nanotubes, nanochannels and mesosponge: Self-organized formation and applications. *Mater. Today* **2013**, *8*, 235–264. [CrossRef]

3. Wierzbicka, E.; Sulka, G.D. Fabrication of highly ordered nanoporous thin Au films and their application for electrochemical determination of epinephrine. *Sens. Actuator B Chem.* **2016**, *222*, 270–279. [CrossRef]

4. Kumeria, T.; Rahman, M.M.; Santos, A.; Ferre-Borrull, J.; Marsal, L.F.; Losic, D. Nanoporous Anodic Alumina Rugate Filters for Sensing of Ionic Mercury: Toward Environmental Point-of-Analysis Systems. *ACS Appl. Mater. Interfaces* **2014**, *6*, 12971–12978. [CrossRef] [PubMed]

5. Stępniowski, W.J.; Salerno, M. Fabrication of nanowires and nanotubes by anodic alumina template assisted electrodeposition. In *Manufacturing Nanostructures*; One Central Press: Altrincham Cheshire, UK, 2014; pp. 321–357.

6. Santos, A.; Yoo, J.H.; Rohatgi, C.V.; Kumeria, T.; Wang, Y.; Losic, D. Realisation and advanced engineering of true optical rugate filters based on nanoporous anodic alumina by sinusoidal pulse anodization. *Nanoscale* **2016**, *8*, 1360–1373. [CrossRef] [PubMed]

7. Santos, A.; Law, C.S.; Pereira, T.; Losic, D. Nanoporous hard data: Optical encoding of information within nanoporous anodic alumina photonic crystals. *Nanoscale* **2016**, *8*, 8091–8100. [CrossRef] [PubMed]

8.   Attauri, A.C.; Huang, Z.; Belwalkar, A.; van Geertruyden, W.; Gao, D.; Misiolek, W. Evaluation of Nano-Porous Alumina Membranes for Hemodialysis Application. *ASAIO J.* **2009**, 217–223. [CrossRef] [PubMed]

9.   Law, C.S.; Santos, A.; Kumeria, T.; Losic, D. Engineered Therapeutic-Releasing Nanoporous Anodic Alumina Aluminum Wires with Extended Release of Therapeutics. *ACS Appl. Mater. Interfaces* **2015**, *7*, 3846–3853. [CrossRef] [PubMed]

10.   Salerno, M.; Loria, P.; Matarazzo, G.; Tome, F.; Diaspro, A.; Eggenhoffner, R. Surface Morphology and Tooth Adhesion of a Novel Nanostructured Dental Restorative Composite. *Materials* **2016**, *9*, 203. [CrossRef] [PubMed]

11.   Feng, X.; Shankar, K.; Paulose, M.; Grimes, C.A. Tantalum-Doped Titanium Dioxide Nanowire Arrays for Dye-Sensitized Solar Cells with High Open-Circuit Voltage. *Angew. Chem. Int. Ed.* **2009**, *121*, 8239–8242. [CrossRef]

12.   Zhang, L.; Liu, L.; Wang, H.; Shen, S.; Chemg, Q.; Yan, C.; Park, S. Electrodeposition of Rhodium Nanowires Arrays and Their Morphology-Dependent Hydrogen Evolution Activity. *Nanomaterials* **2017**, *7*, 103. [CrossRef] [PubMed]

13.   Zhang, X.; Han, F.; Shi, B.; Farsinezhad, S.; Dechaine, G.P.; Shankar, K. Photocatalytic Conversion of Diluted CO_2 into Light Hydrocarbons Using Periodically Modulated Multiwalled Nanotube Arrays. *Angew. Chem. Int. Ed.* **2012**, *51*, 12732–12735. [CrossRef] [PubMed]

14.   Méndez, M.; González, S.; Vega, V.; Teixeira, J.M.; Hernando, B.; Luna, C.; Prida, V.M. Ni-Co Alloy and Multisegmented Ni/Co Nanowire Arrays Modulated in Composition: Structural Characterization and Magnetic Properties. *Crystals* **2017**, *7*, 66. [CrossRef]

15.   Toccafondi, C.; Zaccaria, R.P.; Dante, S.; Salerno, M. Fabrication of Gold-Coated Ultra-Thin Anodic Porous Alumina Substrates for Augmented SERS. *Materials* **2016**, *9*, 403. [CrossRef] [PubMed]

16.   Nesbitt, N.; Merlo, J.M.; Rose, A.H.; Calm, Y.M.; Kempa, K.; Burns, M.J.; Naughton, M.J. Aluminum Nanowire Arrays via Directed Assembly. *Nano Lett.* **2015**, *15*, 7294–7299. [CrossRef] [PubMed]

17.   Kikuchi, T.; Nishinaga, O.; Natsui, S.; Suzuki, R.O. Polymer nanoimprinting using an anodized aluminum mold for structural coloration. *Appl. Surf. Sci.* **2015**, *341*, 19–27. [CrossRef]

18.   Norek, M.; Krasinski, A. Controlling of water wettability by structural and chemical modification of porous anodic alumina (PAA): Towards super-hydrophobic surfaces. *Surf. Coat. Technol.* **2015**, *276*, 464–470. [CrossRef]

19.   Chien, Y.C.; Weng, H.C. A Brief Note on the Magnetowetting of Magnetic Nanofluids on AAO Surfaces. *Nanomaterials* **2018**, *8*, 118. [CrossRef] [PubMed]

20.   Stepniowski, W.J.; Choi, J.; Yoo, H.; Oh, K.; Michalska-Domanska, M.; Chilimoniuk, P.; Czujko, T.; Lyszkowski, R.; Jozwiak, S.; Bojar, Z.; et al. Anodization of FeAl intermetallic alloys for bandgap tunable nanoporous mixed aluminum-iron oxide. *J. Electroanal. Chem.* **2016**, *771*, 37–44. [CrossRef]

21.   Choi, Y.W.; Kim, S.; Seong, M.; Yoo, H.; Choi, J. NH_4-doped anodic WO_3 prepared through anodization and subsequent NH_4OH treatment for water splitting. *Appl. Surf. Sci.* **2015**, *324*, 414–418. [CrossRef]

22.   Kikuchi, T.; Kawashima, J.; Natsui, S.; Suzuki, R.O. Fabrication of porous tungsten oxide via anodizing in an ammonium nitrate/ethylene glycol/water mixture for visible light-driven photocatalyst. *Appl. Surf. Sci.* **2017**, *422*, 130–137. [CrossRef]

23.   Zaraska, L.; Gawlak, K.; Gurgul, M.; Chlebda, D.K.; Socha, R.P.; Sulka, G.D. Controlled synthesis of nanoporous tin oxide layers with various pore diameters and their photoelectrochemical properties. *Electrochim. Acta* **2017**, *254*, 238–245. [CrossRef]

24.   Wierzbicka, E.; Syrek, K.; Sulka, G.D.; Pisarek, M.; Janik-Czachor, M. The effect of foil purity on morphology of anodized nanoporous ZrO_2. *Appl. Surf. Sci.* **2016**, *388*, 799–804. [CrossRef]

25.   Pisarek, M.; Krajczewski, J.; Wierzbicka, E.; Holdynski, M.; Sulka, G.D.; Nowakowski, R.; Kudelski, A.; Janik-Czachor, M. Influence of the silver deposition method on the activity of platforms for chemometric surface-enhanced Raman scattering measurements: Silver films on ZrO_2 nanopore arrays. *Spectrochim. Acta A Mol. Biomol. Spectrosc.* **2017**, *182*, 124–129. [CrossRef] [PubMed]

26.   Park, J.; Kim, K.; Choi, J. Formation of ZnO nanowires during short durations of potentiostatic and galvanostatic anodization. *Curr. Appl. Phys.* **2013**, *7*, 1370–1375. [CrossRef]

27.   Zaraska, L.; Mika, K.; Syrek, K.; Sulka, G.D. Formation of ZnO nanowires during anodic oxidation of zinc in bicarbonate electrolytes. *J. Electroanal. Chem.* **2017**, *801*, 511–520. [CrossRef]

28. Yoo, J.E.; Park, J.; Cha, G.; Choi, J. Micro-length anodic porous niobium oxide for lithium-ion thin film battery applications. *Thin Solid Films* **2013**, *531*, 583–587. [CrossRef]

29. Stojadinovic, S.; Tadic, N.; Radic, N.; Stefanov, P.; Grbic, B.; Vasilic, R. Anodic luminescence, structural, photoluminescent, and photocatalytic properties of anodic oxide films grown on niobium in phosphoric acid. *Appl. Surf. Sci.* **2015**, *355*, 912–920. [CrossRef]

30. Ohta, T.; Masegi, H.; Noda, K. Photocatalytic decomposition of gaseous methanol over anodized iron oxide nanotube arrays in high vacuum. *Mater. Res. Bull.* **2018**, *99*, 367–376. [CrossRef]

31. Stepniowski, W.J.; Choi, J.; Yoo, H.; Michalska-Domanska, M.; Chilimoniuk, P.; Czujko, T. Quantitative fast Fourier transform based arrangement analysis of porous anodic oxide formed by self-organized anodization of FeAl intermetallic alloy. *Mater. Lett.* **2016**, *164*, 176–179. [CrossRef]

32. Zoolfakar, A.S.; Rani, R.A.; Morfa, A.J.; O'Mullane, A.P.; Kalantar-zadeh, K. Nanostructured copper oxide semiconductors: A perspective on materials, synthesis methods and applications. *J. Mater. Chem. C* **2014**, *2*, 5247–5270. [CrossRef]

33. Poulopoulos, P.; Baskoutas, S.; Pappas, S.D.; Garoufalis, C.S.; Droulias, S.A.; Zamani, A.; Kapaklis, V. Intense Quantum Confinement Effects in Cu_2O Thin Films. *J. Phys. Chem. C* **2011**, *115*, 14839–14843. [CrossRef]

34. Musselman, K.P.; Wisnet, A.; Iza, D.C.; Hesse, H.C.; Scheu, C.; MacManus-Driscoll, J.L.; Schmidt-Mende, L. Strong Efficiency Improvements in Ultra-low-Cost Inorganic Nanowire Solar Cells. *Adv. Mater.* **2010**, *22*, E254–E258. [CrossRef] [PubMed]

35. Zhang, Q.; Zhang, K.; Xu, D.; Yang, G.; Huang, H.; Nie, F.; Liu, C.; Yang, S. CuO nanostructures: Synthesis, characterization, growth mechanisms, fundamental properties, and applications. *Prog. Mater. Sci.* **2014**, *60*, 208–337. [CrossRef]

36. Sun, S.; Zhang, X.; Yang, Q.; Liang, S.; Zhang, X.; Yang, Z. Cuprous oxide (Cu_2O) crystals with tailored architectures: A comprehensive review on synthesis, fundamental properties, functional modifications and applications. *Prog. Mater. Sci.* **2018**, *96*, 111–173. [CrossRef]

37. Stepniowski, W.J.; Norek, M.; Budner, B.; Michalska-Domanska, M.; Nowak-Stepniowska, A.; Bombalska, A.; Kaliszewski, M.; Mostek, A.; Thorat, S.; Salerno, M.; et al. In-situ electrochemical doping of nanoporous anodic aluminum oxide with indigo carmine organic dye. *Thin Solid Films* **2016**, *598*, 60–64. [CrossRef]

38. Beverskog, B.; Puigdomenech, I. Revised Pourbaix Diagrams for Copper at 25 to 300 °C. *J. Electrochem. Soc.* **1997**, *144*, 3476–3483. [CrossRef]

39. Gennero de Chiavlo, M.R.; Zerbino, J.O.; Marchiano, S.L.; Arvia, A.J. Correlation of electrochemical and elipsometric data in relation to the kinetics and mechanism of Cu_2O electroformation in alkaline solutions. *J. Appl. Electrochem.* **1986**, *16*, 517–526. [CrossRef]

40. Ambrose, J.; Barrads, R.G.; Shoesmith, D.W. Investigation of copper in aqueous alkaline solutions by cyclic voltammetry. *J. Electronal. Chem. Interfacial Electrochem.* **1973**, *47*, 47–64. [CrossRef]

41. Stepniowski, W.J.; Stojadinovic, S.; Vasilic, R.; Tadic, N.; Karczewski, K.; Abrahami, S.T.; Buijnsters, J.G.; Mol, J.M.C. Morphology and photoluminescence of nanostructured oxides grown by copper passivation in aqueous potassium hydroxide solution. *Mater. Lett.* **2017**, *198*, 89–92. [CrossRef]

42. Allam, N.K.; Grimes, C.A. Electrochemical fabrication of complex copper oxide nanoarchitectures via copper anodization in aqueous and non-aqueous electrolytes. *Mater. Lett.* **2011**, *65*, 1949–1955. [CrossRef]

43. Satheesh Babu, T.G.; Ramachandran, T. Development of highly sensitive non-enzymatic sensor for the selective determination of glucose and fabrication of a working model. *Electrochim. Acta* **2010**, *55*, 1612–1618. [CrossRef]

44. Shooshtari, L.; Mohammadpour, R.; Zad, A.I. Enhanced photoelectrochemical processes by interface engineering, using Cu_2O nanorods. *Mater. Lett.* **2016**, *163*, 81–84. [CrossRef]

45. Cheng, Z.; Ming, D.; Fu, K.; Zhang, N.; Sun, K. pH-Controllable Water Permeation through a Nanostructured Copper Mesh Film. *ACS Appl. Mater. Interfaces* **2012**, *4*, 5826–5832. [CrossRef] [PubMed]

46. Xiao, F.; Yuan, S.; Liang, B.; Li, G.; Pehkonen, S.O.; Zhang, T.J. Superhydrophobic CuO nanoneedle-covered copper surfaces for anticorrosion. *J. Mater. Chem. A* **2015**, *3*, 4374–4388. [CrossRef]

47. Wu, X.; Bai, H.; Zhang, J.; Chen, F.; Shi, G. Copper hydroxide nanoneedle and nanotube arrays fabricated by anodization of copper. *J. Phys. Chem. B* **2005**, *109*, 22836–22842. [CrossRef] [PubMed]

48. Wang, P.; Ng, Y.H.; Amal, R. Embedment of anodized p-type Cu_2O thin films with CuO nanowires for improvement in photoelectrochemical stability. *Nanoscale* **2013**, *5*, 2952–2958. [CrossRef] [PubMed]

49. Oyarzún Jerez, D.P.; López Teijelo, M.; Ramos Cervantes, W.; Linarez Pérez, O.E.; Sánchez, J.; Pizarro, G.C.; Acosta, G.; Flores, M.; Arratia-Perez, R. Nanostructuring of anodic copper oxides in fluoride-containing ethylene glycol media. *J. Electroanal. Chem.* **2017**, *807*, 181–186. [CrossRef]

50. Wu, J.; Li, X.; Yadian, B.; Liu, H.; Chun, S.; Zhang, B.; Zhou, K.; Gan, C.L.; Huang, Y. Nano-scale oxidation of copper in aqueous solution. *Electrochem. Commun.* **2013**, *26*, 21–24. [CrossRef]

51. Caballero-Briones, F.; Palacios-Padrós, A.; Calzadilla, O.; Sanz, F. Evidence and analysis of parallel growth mechanisms in Cu₂O films prepared by Cu anodization. *Electrochim. Acta* **2010**, *55*, 4353–4358. [CrossRef]

52. Jiang, W.; He, J.; Xiao, F.; Yuan, S.; Lu, H.; Liang, B. Preparation and Antiscaling Application of Superhydrophobic Anodized CuO Nanowire Surfaces. *Ind. Eng. Chem. Res.* **2015**, *54*, 6874–6883. [CrossRef]

53. Wan, Y.; Zhang, Y.; Wang, X.; Wang, Q. Electrochemical formation and reduction of copper oxide nanostructures in alkaline media. *Electrochem. Commun.* **2013**, *36*, 99–102. [CrossRef]

54. Zhang, Z.; Wang, P. Highly stable copper oxide composite as an effective photocathode for water splitting via a facile electrochemical synthesis strategy. *J. Mater. Chem.* **2012**, *22*, 2456–2464. [CrossRef]

55. Caballero-Briones, F.; Artes, J.M.; Díez-Perez, I.; Gorostiza, P.; Sanz, F. Direct Observation of the Valence Band Edge by in Situ ECSTM-ECTS in p-Type Cu₂O Layers Prepared by Copper Anodization. *J. Phys. Chem. C* **2009**, *113*, 1028–1036. [CrossRef]

56. Shu, X.; Zheng, H.; Xu, G.; Zhao, J.; Cui, L.; Qin, Y.; Wang, Y.; Zhang, Y.; Wu, Y. The anodization synthesis of copper oxide nanosheet arrays and their photoelectrochemical properties. *Appl. Surf. Sci.* **2017**, *412*, 505–516. [CrossRef]

57. Arurault, L.; Belghith, M.H.; Bes, R.S. Manganese pigmented anodized copper as solar selective absorber. *J. Mater. Sci.* **2007**, *42*, 1190–1195. [CrossRef]

58. Zhang, Z.; Zhong, C.; Liu, L.; Teng, X.; Wu, Y.; Hu, W. Electrochemically prepared cuprous oxide film for photo-catalytic oxygen evolution from water oxidation under visible light. *Sol. Energy Mater. Sol. Cells* **2015**, *132*, 275–281. [CrossRef]

59. Khan, B.S.; Saeed, A.; Hayat, S.S.; Mukhtar, A.; Mehmood, T. Mechanism for the Formation of Cuprous Oxide Nanowires in AAO template by Electrodeposition. *Int. J. Electrochem. Sci.* **2017**, *12*, 890–987. [CrossRef]

60. Akhavan, O.; Tohidi, H.; Moshfegh, A.Z. Synthesis and electrochromic study of sol-gel cuprous oxide nanoparticles accumulated on silica thin film. *Thin Solid Films* **2009**, *517*, 6700–6706. [CrossRef]

MDPI
St. Alban-Anlage 66
4052 Basel
Switzerland
Tel. +41 61 683 77 34
Fax +41 61 302 89 18
www.mdpi.com

Nanomaterials Editorial Office
E-mail: nanomaterials@mdpi.com
www.mdpi.com/journal/nanomaterials

MDPI
St. Alban-Anlage 66
4052 Basel
Switzerland

Tel: +41 61 683 77 34
Fax: +41 61 302 89 18

www.mdpi.com

MDPI

ISBN 978-3-03897-269-3

www.ingramcontent.com/pod-product-compliance
Lightning Source LLC
Chambersburg PA
CBHW041218220326
41597CB00033BA/6012